# Chemistry 11 In Focus

## Solution Manual

The New Awakening Books—Madison, WI
ISBN: 979-8-218-14128-8
Library of Congress Control Number: pending
Title: *Chemistry 11 In Focus Solution Manual*
Author: Abdul Jalil Shakur
Digital distribution | 2023
Paperback | 2023

**Editors:**
Mohamed Kazim Yusuff
**C.P.A., C.M.A.**
Clive Sankardyal
**BA. Dip.ED**
Dr. Yaseer Shakur
**BSc., MSc.,PhD., MD.**

**Revisers:**
Dr. Yaseer Shakur
**BSc., MSc.,PhD., MD.**

**Cover Page:**
Nafeesa Shakur

**Cover Molecule:**
Abdul Shakur

# Dedication

I would like to dedicate this book to all those students who encounter difficulties in Grade 11 Chemistry, for various reasons. Also, to all sick, poor and oppressed people around the world. And last, but not least, to all members of my beloved family.

# Table of Contents


Unit 1: Matter and Chemical Bonding
Chapter 1: Classification of matter (Exercise 1.1)..............................................................................1
Chapter 2: Atomic theories (Exercise 2.1)........................................................................................1
Chapter 3: Isotopes – half-life problems (Exercise 3.1) ....................................................................1
    Balancing nuclear equations (Exercise 3.2).................................................................................2
    Half-life problems (Exercise 3.3) ...............................................................................................2
Chapter 4: Trends in the periodic table – ionization energy (Exercise 4.1) .......................................2
Chapter Review: Matter (Chapters 1-4)............................................................................................4
Chapter 5: Chemical Bonding
    Bohr-Rutherford diagram for ions (Exercise 5.1).....................................................................11
    Bohr-Rutherford diagram for ions (Exercise 5.2).....................................................................11
    Bohr-Rutherford diagrams for ionic compounds (Exercise 5.3) ...............................................12
    Lewis dot diagram for ionic compounds (Exercise 5.4)............................................................13
    Lewis dot diagram for molecular compounds (Exercise 5.5).....................................................13
Chapter 6: Molecular Geometry
    Polarity of covalent bonds (Exercise 6.1).................................................................................14
    Polarity of covalent bonds (Exercise 6.2).................................................................................14
    Molecular geometry (Exercise 6.3) ..........................................................................................14
    Properties of molecular compounds (Exercise 6.4) ...................................................................14
    Properties of molecular compounds (Exercise 6.5) ...................................................................15
    Polarity of molecular compounds (Exercise 6.6).......................................................................16
Chapter Review: Chemical Bonding (5-6) ......................................................................................17
Chapter 7:  Chemical Reactions
    Evidence for chemical changes (Exercise 7.1) ..........................................................................23
    Balancing chemical equations from word equations (Exercise 7.2)............................................23
    Balancing chemical equations from chemical formulas (Exercise 7.3)........................................23
    Types of Chemical Reactions (Exercise 7.4).............................................................................24
    Balancing single displacement reactions (Exercise 7.5).............................................................24
    Relative reactivity of metals (Exercise 7.6)...............................................................................25
    Double Displacement Reactions (Exercise 7.7) ........................................................................25
    Balancing double displacement reactions (Exercise 7.8) ...........................................................26
    Combustion of Acetylene (Analysis)........................................................................................26
    Flame tests (Exercise 7.9).......................................................................................................27
Chapter 8: Names and Formulas for Chemical Compounds
    Writing formulas for binary ionic compounds (Exercise 8.1) ....................................................28
    Writing names for binary ionic compounds (Exercise 8.2)........................................................28
    Writing names for binary molecular compounds (Exercise 8.3)..................................................28
    Writing formulas for binary molecular compounds (Exercise 8.4) .............................................28
    Writing IUPAC names for binary compounds (Exercise 8.5) ....................................................29
Chapter Review: Reactions (Chapters7-8) .....................................................................................30

Unit 2: Quantities in Chemical Reactions
Chapter 9: The mole concept
    Review of atomic notation (Exercise 9.1) .................................................................................38

Review of atomic notation (Exercise 9.2)........................................................................38
The mole concept (Exercise 9.3) .....................................................................................38
Finding masses of elements, given moles and molar masses (Exercise 9.4).....................39
Finding number of moles of atoms, given their masses and molar masses (Exercise 9.5)..................40
Calculating number of entities (Exercise 9.6)..................................................................41
Converting amounts in mass to number of moles (Exercise 9.7) .....................................41
Converting number of entities to number of moles (Exercise 9.8)...................................42
Converting number of entities to amount in mass (Exercise 9.9)....................................43
Distinction between ionic and molecular particles (Exercise 9.10)..................................44
Finding molar masses of compounds (Exercise 9.11) ......................................................44

## Chapter 10: Composition of compounds

Percentage composition of compounds (Exercise 10.1) .....................................................47
Finding empirical and molecular formulas of compound (Exercise 10.2) ........................49

## Chapter 11: Quantities in Chemical reactions

Introduction to stoichiometry (Exercise 11.1) ................................................................57
Quantities in chemical reactions (Exercise 11.2)..............................................................57
Excess and limiting reagents (Exercise 11.3) ...................................................................61
Calculating Percentage yield (Exercise 11.4) ...................................................................65
Chapter Review: Quantities in chemical reactions (9-11) ................................................77

# Unit 3: Solutions and Solubility

## Chapter 12: Nature and Properties of Solutions

Dissociation equations for ionic compounds (Exercise 12.1)..............................................85
Homogeneous versus heterogenous solutions (Exercise 12.2) ..........................................85

## Chapter 13: Concentration of solutions

Molar concentration of solution (Exercise 13.1) ..............................................................86
Molar concentration of solutions (Exercise 13.2)..............................................................88
Calculating masses for solutions preparation (Exercise 13.3)...........................................91
Preparations of solutions from hydrated salts (Exercise 13.4) .........................................94
Dissociation of ionic compounds in water (Exercise 13.5) ...............................................95
Ionization of molecular compounds in water (Exercise 13.6) ...........................................95
Calculation molar concentration of ions (Exercise 13.7)..................................................96
Problems on dilution of solutions (Exercise 13.8).............................................................99
Percentage concentration of compounds (Exercise 13.9) ................................................105
Very low concentrations (Exercise 13.11)......................................................................108
Molar concentrations from density and percentage by volume (Exercise 13.12)..............109

## Chapter 14: Factors that affect solubility of substances in water

Solubility of Gases activity (Analysis 1.) .......................................................................110
Solubility Curves (Analysis 2.)......................................................................................110
Solubility Curves (Analysis 3.)......................................................................................110
Supersaturated solution (Exercise 14.1) ........................................................................111

## Chapter 15: Soaps and Detergents

Investigating hardness of water (Analysis 1.)..................................................................111
Causes of hardness of water (Analysis 2.) ......................................................................112
Softening of hard water (Analysis 3.).............................................................................112
Effects of hard water on detergents (Analysis 4.)............................................................113
Chapter Review: Solution and solubility .......................................................................114
Chapter 16: Sewage treatment and water purification
Canadian guidelines for drinking water quality (Exercise 16.1) .......................................119
Effects of Biodegradable organic matter on dissolved gases (Analysis 4.)........................119

Chapter Review: Sewage treatment and water treatment ......................................................................121
Chapter 17: Acids and Bases
    The Bronsted-Lowry theory (Exercise 17.1) ...........................................................................125
    Acids and Base strength (Exercise 17.3) ................................................................................125
    Balancing acid-base neutralization equations (Exercise 17.4) ..........................................125
    Solving acid-base neutralization problems (Exercise 17.5)................................................126
    Solving acid-base titration problems (Exercise 17.6)..........................................................137
    Finding pH of solutions, given concentration of $H^+_{(aq)}$ (Exercise 17.7) ...........................142
    Finding pH of solutions, given mass and volume of acids (Exercise 17.8)......................143
    Finding $H^+_{(aq)}$ concentrations from pH (Exercise 17.9) ......................................................146
    Finding the resulting pH from acid-base mixtures (Exercise 17.11) .................................147
    Writing Formulas for polyatomic ionic compounds (Exercise 17.12) ..............................158
    Writing Names for polyatomic ionic compounds (Exercise 17.13) ..................................158
Chapter Review: Acids and Bases........................................................................................................159

Unit 4: Gases
Chapter 18: The Gas Laws
    Boyle's Law problems (Exercise 18.1).....................................................................................165
    Temperature Conversions (Exercise 18.2)..............................................................................167
    Charles' Law problems (Exercise 18.3) ...................................................................................167
    Temperature and Pressure law problems (Exercise 18.4)....................................................169
    The Combined Gas Law problems (Exercise 18.5)................................................................171
Chapter 19: Gas Mixtures
    Dalton's Law of partial pressure problems (Exercise 19.1) ................................................175
    Molar volumes problems (Exercise 19.2)................................................................................175
Chapter 20: The ideal gas law
    The ideal gas law problems (Exercise 20.1) ...........................................................................180
Chapter 21: Atmospheric Pollution
    The nitrogen Cycle: (Exercise 20.11) ......................................................................................186
Chapter Review: (18-21) .........................................................................................................................188
The periodic Table....................................................................................................................................194

# UNIT 1: Matter and Chemical Bonding

## CHAPTER 1: Classification of matter

**Exercise 1.1 (p. 6-7)**

1. (a) liquid  (b) solid  (c) gas  (d) liquid  (e) solid
2. (a) sublimation (b) evaporation (c) melting (d) condensation (e) freezing
3. (a) physical change  (b) physical property  (c) chemical property
   (d) chemical change  (e) chemical property  (f) physical change
   (g) physical property
4. (a) pure  (b) heterogeneous mixture  (c) homogeneous mixture
   (d) heterogeneous mixture  (e) homogeneous mixture  (f) pure
5. (a) released  (b) released  (c) absorbed (d) released  (e) absorbed
   (f) absorbed
6. (a) pure  (b) quantitative  (c) weakest (d) qualitative  (e) impure
   (f) impure (g) pure
7. (a) false  (b) false  (c) true  (d) false  (e) false  (f) true  (g) true

## CHAPTER 2 : Atomic theories

**Exercise 2.1 (p.18)**

1. (a) Chadwick (b) Thomson (c) Goldstein (d) Rutherford (e) Bohr (f) Planck
   (g) Dalton (h) Nagaoka.
2. (a) positively (b) electrons (c) protons (d) neutrons (e) shells (f) energy levels
   (g) force (h) positively (i) ground (j) quantum (k) jump (l) excited
   (m) released (n) light (o) line.
3. (a) potassium nitrate, carbon in the form of charcoal, and sulfur, in a %75: %15 %10 ratio
   (b) At a high temperature, the potassium nitrate decomposes to produce oxygen that is required for the reaction to take place.
   (c) The heat causes the metal atoms or their ions to become excited (their electrons being bumped to higher energy levels) and when the electrons return to their ground state, they release energy in the form of light. The color of light produced is specific to each metal. A combination of different metal salts will therefore produce a series of different colors, depending on which ones are used.
   (d) Salts containing lithium, copper, and barium ions.
4. Since UV has longer wavelength than X-rays, it is less penetrating than X-rays; sunscreens which are made up of mixtures of titanium dioxide and zinc are thus used to block it. Lead shields of certain thickness are used to block the more penetrative X-rays from producing mutations in our gonads. The patients wear these as aprons when they take X-ray scans.

## CHAPTER 3: Isotopes

**Exercise 3.1 (p.23)**

**Answer to graph on Iodine**
1. 8.07 days  **2.** 24.2 days  **3.** 32.3 days  **4.** The first  **5.** 16.1 days

1. (a) $^{14}_{6}C \longrightarrow {}^{14}_{7}C + {}^{0}_{-1}e$  - beta decay
   (b) $^{31}_{15}P \longrightarrow {}^{28}_{14}Si + {}^{3}_{1}H$   - nuclear fission
   (c) $^{90}_{38}Sr \longrightarrow {}^{90}_{39}Sr + {}^{0}_{-1}e$   - beta decay
   (d) $^{73}_{31}Ga \longrightarrow {}^{0}_{-1}e + {}^{73}_{32}Ga$  - beta decay
   (e) $^{231}_{90}Th + {}^{1}_{0}n \longrightarrow {}^{232}_{90}Th$   - neutron enrichment
   (f) $^{12}_{5}B \longrightarrow {}^{7}_{3}Li + {}^{1}_{0}n + {}^{4}_{2}He$  - nuclear fission
   (g) $^{226}_{88}Ra \longrightarrow {}^{4}_{2}He + {}^{222}_{86}Rn$   - nuclear fission



**Answer to graph on Antibiotic:**

1. (a) 160 days   (b) 80 days   (c) 140 g
2. One with a short half-life. Since an acute illness lasts for a relatively short time, an antibiotic with a short half-life is the better choice as it would disappear from the body quicker with minimum side effects.
3. 28.09,  4. K-41       6. 0.014 Bq

## Chapter 4 : Trends in the periodic table



1. a) **The nuclear charge**. As the nuclear charge increases it tends to decrease the atomic radius by attracting the electron shells closer to it.

   However, as the number of electron shells increases, the radius increases so exponentially that it negates any significant decrease of atomic radius caused by an increase of nuclear charge.

2. It decreases down the group. As you go down the group, the number of electron shells increases by one, making the atomic radius increase. The outer electron is attracted less to the nuclear charge because it is much farther away from the nucleus. Also, as the number of inner electron shells increases, they shield the outer electron from the nuclear charge, making it easier for it to leave the atom.

3. The first ionization energy is less than the second, because electron-electron repulsion was involved in pushing the first of the two electrons off the atom. After the first of the two electrons is ejected, the second one does not experience any electron-electron repulsion; more energy is thus required to eject it off the remaining positive ion.

4. When the fluoride atom gains an electron, there is now more electron-electron repulsion that causes expansion of the valence shell with a greater radius.

5.

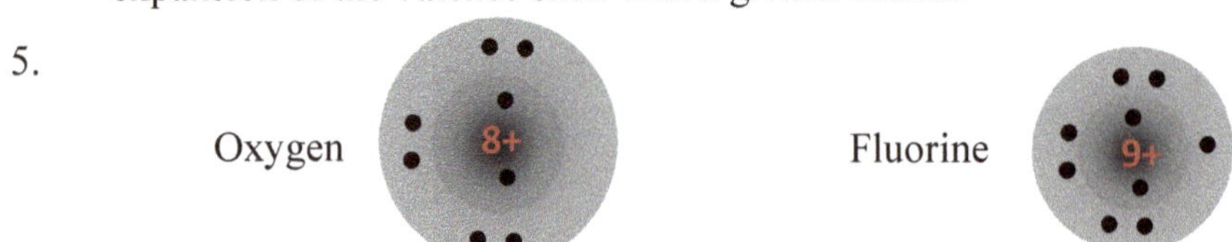


Both oxygen and fluorine have the same number of electron shells. However, since fluorine has a greater effective nuclear charge than oxygen, its nucleus exerts a greater pull on its outer electron shell than the nucleus of oxygen does to its outer electrons. The atomic radius of fluorine is thus smaller than oxygen. Since the outer electrons of fluorine are attracted with a greater force of attraction than those of oxygen, the ionization energy fluorine will be greater than that of oxygen.

6. Magnesium has a smaller atomic radius because it has a greater nuclear charge than sodium; the force of attraction between the nucleus of magnesium and its outer shell electrons is greater than the force of attraction between the nucleus of sodium and its outer electrons. It is more difficult to remove a valence electron from the magnesium atom than it is from a sodium atom; the ionization energy of magnesium is thus larger than that of sodium.

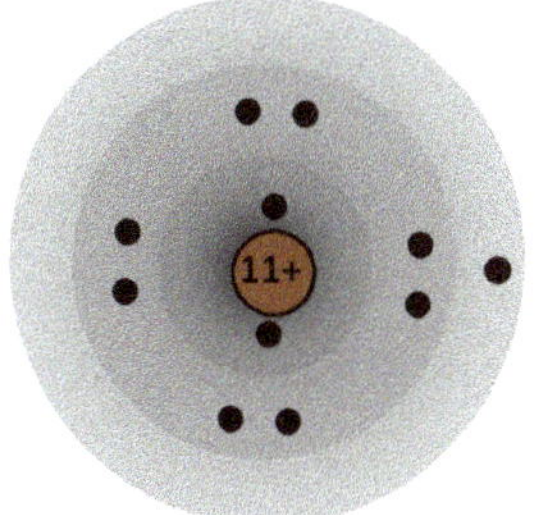

Sodium atom

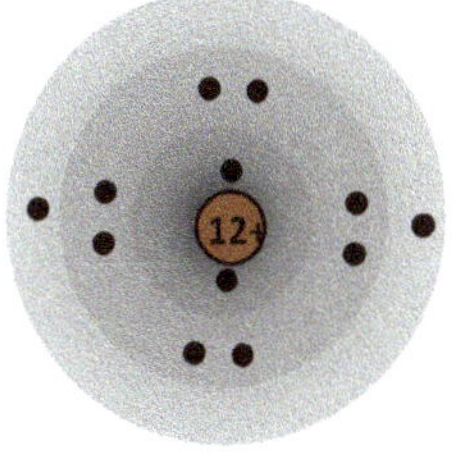

Magnesium atom

# Chapter Review: Matter 1-4  (p.37-41) Solution

## Matching

*Match each term in the table below with the correct statements that follow:*

| | | | |
|---|---|---|---|
| A | Half-life | G | Isotopes |
| C | Thomson | H | Effective nuclear charge |
| B | Ionization energy | I | Alpha particles |
| D | Bohr | J | Chadwick |
| E | Transition | L | $E = hv$ |
| F | Electron affinity | M | $4\,{}^{1}_{1}H + 2\,{}^{0}_{-1}e \rightarrow {}^{4}_{2}He + Energy$ |

1. As this increases the atomic radius decreases.

2. The amount of energy required to the outermost electrons from an atom.

3. The energy change associated when an electron is added to an atom in its gaseous state.

4. He calculated the charge-to-mass ratio of the electron.

5. These were used by Rutherford to probe the atoms of a gold foil.

6. He discovered the neutron.

7. He proposed that electrons are found in energy levels around the nucleus of atoms.

8. The equation for the amount of energy of a photon.

9. This describes the movement of an electron from a lower energy level to a higher energy level.

10. This equation represents a nuclear fusion reaction.

11. The time it takes for half the amount of radioactive atoms in a sample to decay.

12. The nuclei of these have the same number of protons but different number of neutrons.

## True or False

**Read each of the following statements and then decide if it is *True* or *False*.**

1. All compounds are pure substances.

2. Milk is a homogeneous mixture.

3. The sublimation of $CO_{2(s)}$ is chemical change.

4. Viscosity is an example of a physical property.

5. The second ionization energy is always greater than the first.

6. The electron affinity of chlorine is greater than that of fluorine.

7. Cathode rays are negatively charged.

8. Beta radiation is more energetic than gamma radiation.

9. Gamma radiation is electromagnetic in nature.

10. An atom in its excited state is more stable than in its ground state.

11. Carbon has a smaller atomic radius than boron.

12. To break the octet requires a large amount of energy.

13. When an atom absorbs an electron, energy is always released.

14. The mass of one proton has the exact mass as that of a neutron.


## Multiple Choice

**Choose the letter that best answers the questions:**

1. Both oxygen and hydrogen are classified as pure substances because they each contain only one type atoms. Water is also classified as a pure substance even though it is made up of both hydrogen and oxygen atoms. This so because

| | |
|---|---|
| a | both oxygen and hydrogen are pure substances |
| b | water is composed of only one type of particles, which in this case are molecules |
| c | oxygen and water molecules are chemically combined to form only water molecules |
| d | hydrogen and water are physically combined to form water molecules |
| e | b and c |

2. Potassium explodes when added to water. This change can be classified as

| | | | |
|---|---|---|---|
| a | physical | d | evaporation |
| b | sublimation | e | none of the above |
| c | chemical | | |

3. Cathode rays were reclassified as electrons because

| | |
|---|---|
| a | they originated from the cathode and travelled towards the anode |
| b | they were repelled by the negative electrode and attracted by the positive electrode |
| c | they were deflected by a magnetic field placed perpendicular to their path. |
| d | they had the same charge to mass ratio when produced from different cathode materials |
| e | all of the above |

4.

Isotopes of an element are due to atoms of that element having

| a | the same number of protons but different number of neutrons |
| b | the same number of protons but different number of electrons |
| c | the same number of protons and neutrons |
| d | the same number of neutrons, but a different number of protons |
| e | different number of protons and neutrons |

5.

Since the relative atomic mass of Zinc is (65.41 u), it means that

| a | the nuclei of zinc have fractions of nucleons (protons and neutron) |
| b | the element zinc is composed of a mixture of isotopes. |
| c | the nuclei zinc atoms are broken |
| d | zinc atoms have unstable nuclei |
| e | zinc atoms are radioactive. |

6.

For an emission spectrum to be **visible** in any excited atom, electrons must fall back to an energy level of

| a | principal quantum number, n = 1 | d | principal quantum number, n = 4 |
| b | principal quantum number, n = 2 | e | principal quantum number, n = 5 |
| c | principal quantum number, n = 3 | | |

7.

Nitrogen has a smaller atomic radius than beryllium because

| a | a nitrogen atom is lighter than a beryllium atom |
| b | a nitrogen atom has lesser number of electronic shells than beryllium atom |
| c | a nitrogen atom has greater effective nuclear charge than a beryllium atom |
| d | a nitrogen atom is in a different period than a beryllium |

8.

The half-life of plutonium-239 is 24 100 years. If 100 g sample of it undergoes radioactive decay for 48 200 years, what mass of it would remain?

| a | 25 g | d | 12.5 g |
| b | 50 g | e | None of the above |
| c | 75 g | | |

9.  Potassium has a bigger atomic radius than lithium because

| | |
|---|---|
| a | as you go down a group, the atomic radius increases |
| b | potassium has more protons in its nucleus than beryllium |
| c | potassium weighs more than lithium |
| d | potassium has more electronic shells than lithium |
| e | a and d |

10. For the following nuclear equation: $^{226}_{88}Ra \longrightarrow \quad ? \quad + \quad ^{4}_{2}He$

The equation above can be completed by which of the following elements:

| | | | |
|---|---|---|---|
| a | $^{222}_{86}Rn$ | d | $^{230}_{90}Ra$ |
| b | $^{209}_{84}Po$ | e | $^{227}_{89}Ac$ |
| c | $^{210}_{85}At$ | | |

11. The atomic mass of any atom is primarily due to its number of

| | | | |
|---|---|---|---|
| a | protons and electrons in the atom | d | neutrons and electrons in the atom |
| b | electrons in the atom | e | protons in the nucleus |
| c | protons and neutrons in the nucleus | | |

12. The ionic radius of a cation is smaller than its corresponding atomic radius because

| | |
|---|---|
| a | the cation has a greater effective nuclear charge than the corresponding atom |
| b | the atom has a greater effective nuclear charge than the corresponding cation |
| c | the cation may have less energy levels than the corresponding atom |
| d | the cation has less electron-electron repulsion in its valence shell than the corresponding atom |
| e | both c and d |

13. The ionic radius of an anion is greater than its corresponding atomic radius because

| | |
|---|---|
| a | the anion has a greater effective nuclear charge than the corresponding atom |
| b | the atom has a greater effective nuclear charge than the corresponding anion |
| c | the anion has more energy levels than the corresponding atom |
| d | the atom has more energy levels than the corresponding anion |
| e | the anion has a greater electron-electron repulsion in its valence shell than the corresponding atom |

**14.** The second ionization energy of magnesium is greater than its first because

| | |
|---|---|
| *a* | the ionic radius is bigger than the atomic radius |
| *b* | there is now a greater effective nuclear charge |
| *c* | there is now less electron-electron repulsion in its valence shell |
| *d* | there is now more electron-electron repulsion in its valence shell |
| *e* | none of the above |

**15.** Groups IIA elements have positive values for electron affinity because

| | |
|---|---|
| *a* | the change is endothermic |
| *b* | the added electron(s) goes to a subshell that is farther away from the nucleus |
| *c* | energy is liberated |
| *d* | no energy change happens |
| *e* | a and b |

**16.** Nitrogen has a greater first ionization energy than carbon because

| | |
|---|---|
| *a* | they are both in the same period, but nitrogen has a smaller atomic radius than carbon |
| *b* | carbon has a smaller effective nuclear charge than nitrogen |
| *c* | nitrogen needs more energy to remove its electron |
| *d* | all of the above |
| *e* | none of the above |

**17.** The energy of a photon of light is dependent on

| | | | | |
|---|---|---|---|---|
| *a* | its frequency | *d* | its source |
| *b* | its intensity | *e* | a and c |
| *c* | wavelength | | |

**18.** Hydrogen ions from dilute acids accept electrons from dilute acids and liberates hydrogen according to the following equation: $2\,H^+_{(aq)} + 2\,e^- \longrightarrow H_{2(g)}$. Three metals X, Y and Z have the following first ionization energies respectively: $E_1$ = 906, 899l and 738 (kJ/mol). Which of these two metals will liberate hydrogen at the faster rate from dilute acids?

| | | | | |
|---|---|---|---|---|
| *a* | X | *d* | not enough information given |
| *b* | Y | *e* | None of the above |
| *c* | Z | | |

19.

An element has the following four successive ionization energies in kJ/mol.

| 1st | 2nd | 3rd | 4th |
|---|---|---|---|
| 801 | 2 427 | 3 660 | 25 026 |

The element is most likely to be a

| | | | |
|---|---|---|---|
| a | an inert gas | d | group 4 element |
| b | group 2 element | e | group 1 element |
| c | group 3 element | | |

20. The atomic mass unit (amu) can be which of the following?

| | | | |
|---|---|---|---|
| a | the mass of a proton | d | all of the above |
| b | the mass of a neutron | e | the mass of an electron |
| c | The mass of one twelfth the mass of a C-12 atom | | |

21. Bromine has a bigger atomic radius than chlorine because

| | |
|---|---|
| a | bromine is less reactive than chlorine |
| b | bromine has more protons in its nucleus than beryllium |
| c | bromine weighs more than chlorine |
| d | bromine has more electronic shells than chlorine |
| e | none of the above |

22. Which of the following sub-shells does the K shell have?

| | |
|---|---|
| a | s alone |
| b | p alone |
| c | both s and p |
| d | s, p and d |
| e | none of the above |

# Chapter Review: Matter (1-4) Solution

**Matching :**

| | | | | | | | |
|---|---|---|---|---|---|---|---|
| 1.) H | 2.) B | 3.) F | 4.) C | 5.) I | 6.) J | 7.) D | 8.) L |
| 9.) E | 10.) M | 11.) A | 12.) G | | | | |

**True/False**

| | | | | | | | |
|---|---|---|---|---|---|---|---|
| 1.) T | 2.) F | 3.) F | 4.) T | 5.) F | 6.) F | 7.) F | 8.) F |
| 9.) T | 10.) F | 11.) T | | | | | |

**Multiple Choice**

| | | | | | | | |
|---|---|---|---|---|---|---|---|
| 1.) E | 2.) C | 3.) E | 4.) A | 5.) B | 6.) B | 7.) C | 8.) A |
| 9.) E | 10.) A | 11.) C | 12.) E | 13.) E | 14.) C | 15.) E | 16.) D |
| 17.) E | 18.) C | 19.) C | 20.) D | 21.) D | 22.) C | | |

# Chapter 5 : Chemical Bonding



(a)

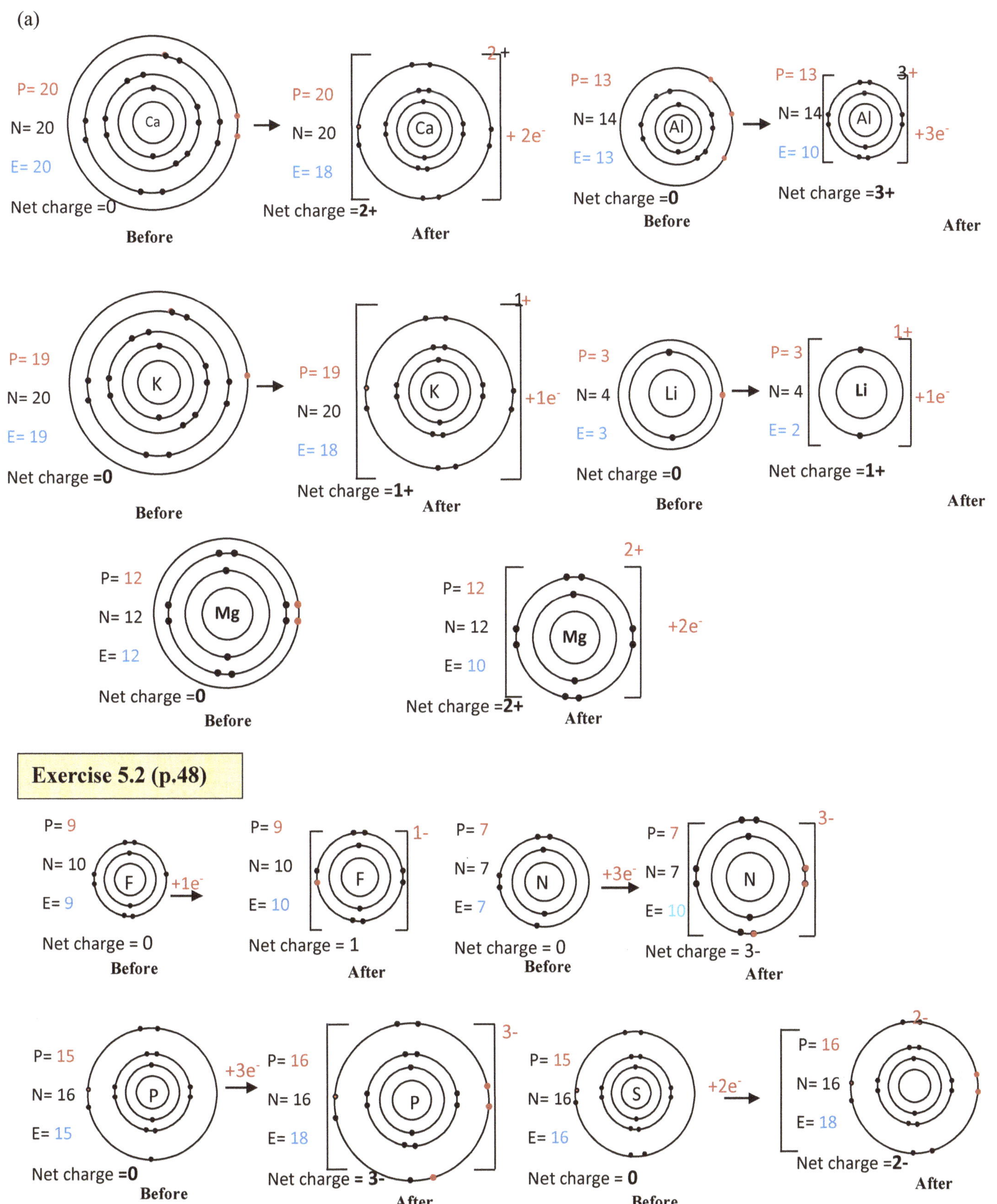

Exercise 5.3 (p.49)


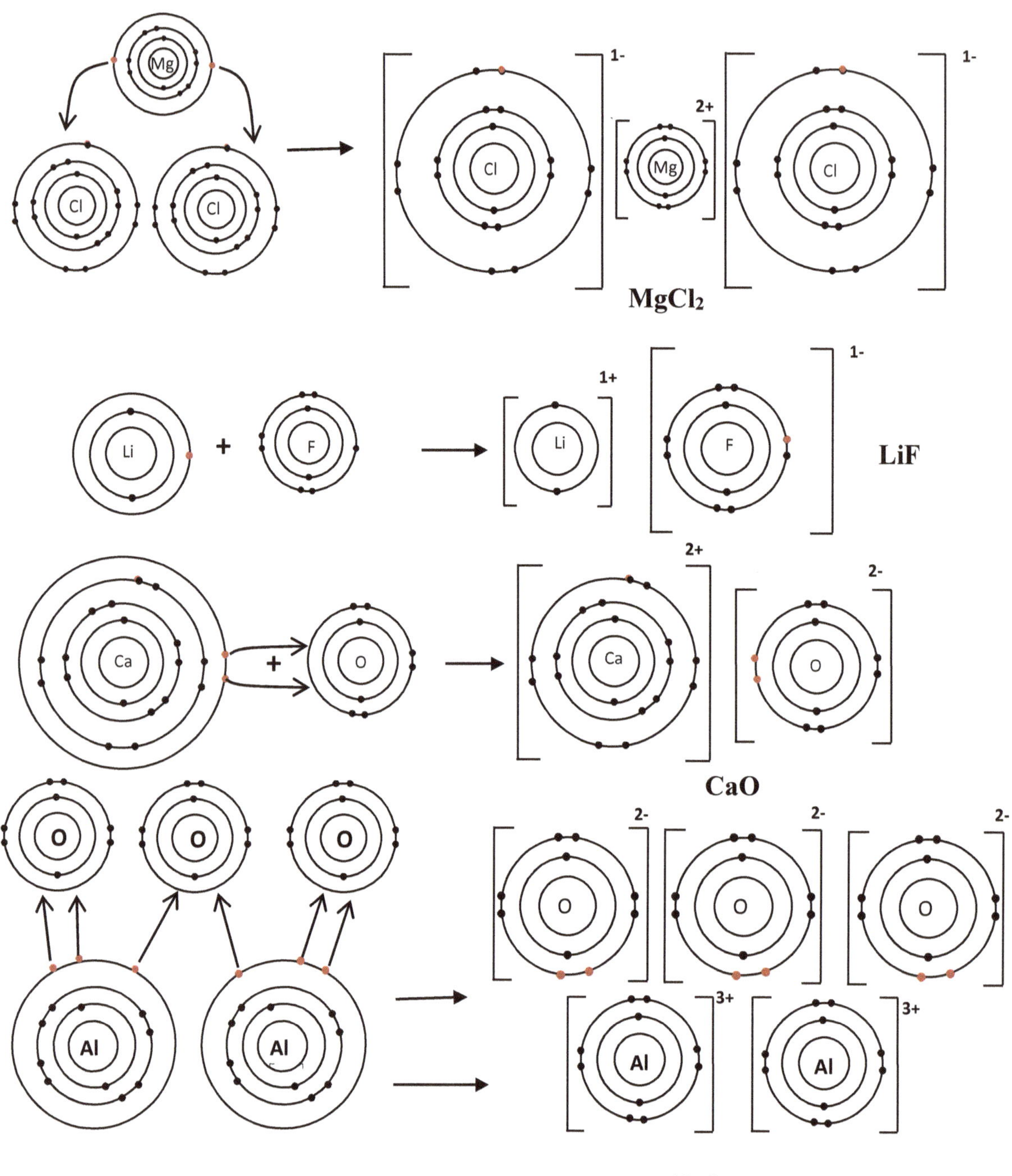

Mg
Cl
Cl
Cl
Mg
Cl
MgCl₂
Li
F
Li
F
LiF
Ca
O
Ca
O
CaO
O
O
O
O
O
O
Al
Al
Al
Al
Al₂O₃

**Exercise 5.4 (p.50)**

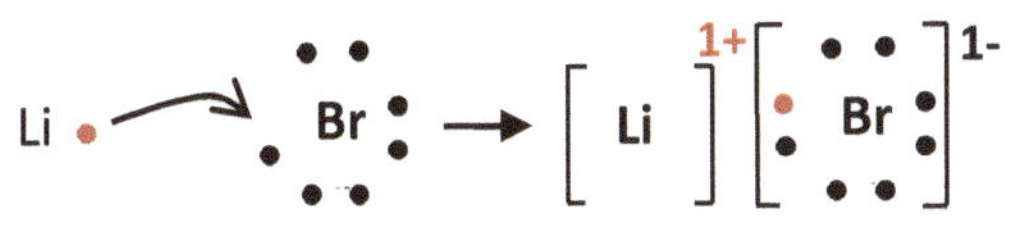

LiBr

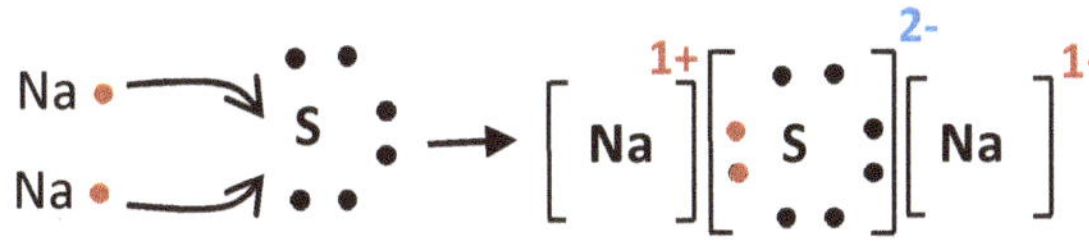

$Na_2S$

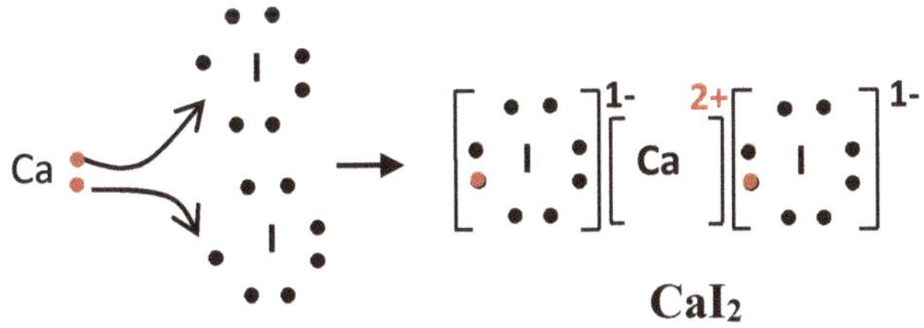

$CaI_2$

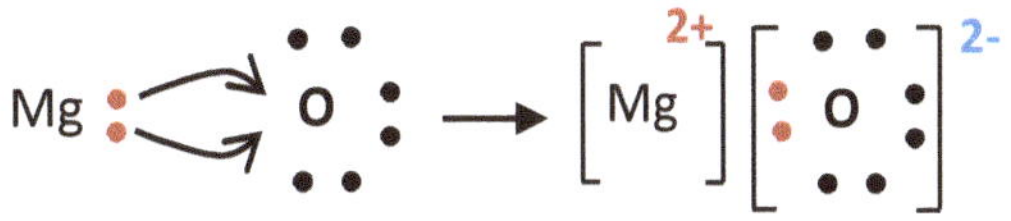

MgO

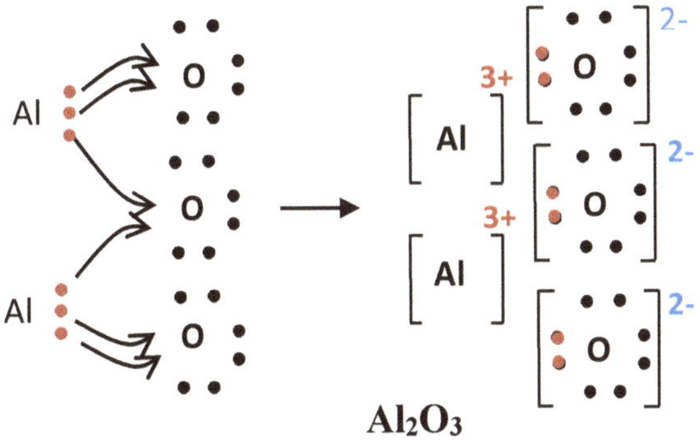

$Al_2O_3$

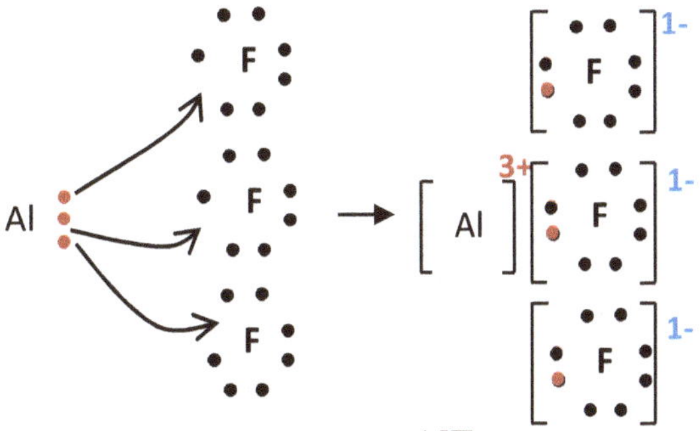

$AlF_3$

**Exercise 5.5 (p.61)**

(a) 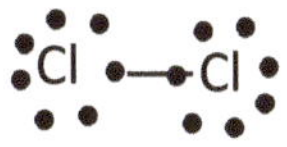

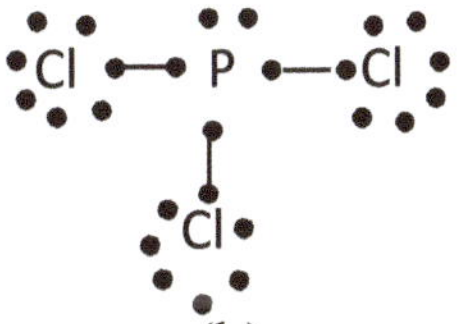

(b)

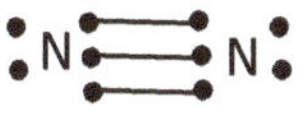

(c)

(d) 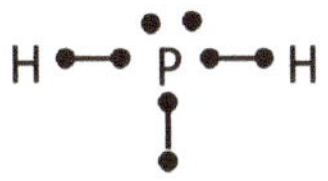

(e) 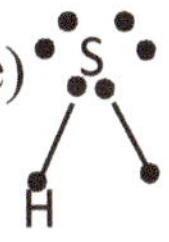

H

(f) 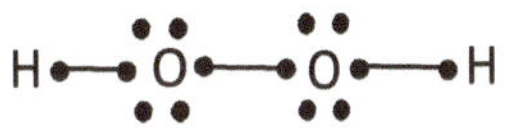

(g) 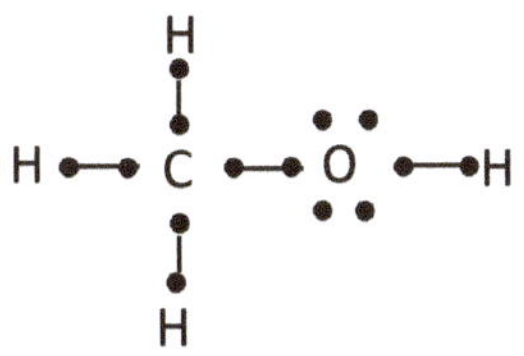

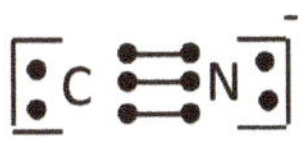

(h)

(i) 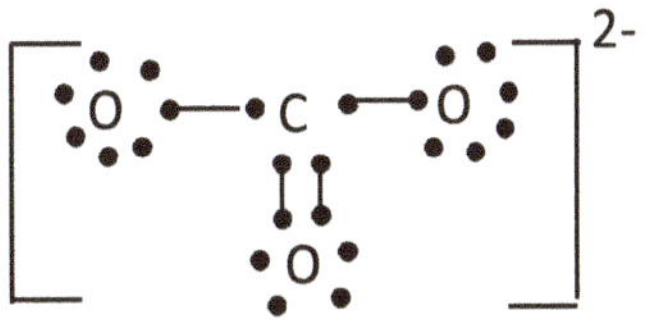

* *Please ignore the dots at the ends as they are inserted only as a guide to the electron count.*

# Chapter 6: Molecular geometry

**Exercise 6.1 (p.64)**

(a) polar   (b) polar  (c) polar  (d) polar  (e) polar  (f) polar

**Exercise 6.2 (p.48)**

(a) polar        (b) polar      (c)   polar       (d) polar

**Exercise 6.3 (p.74)**

**Table 6.3:** Complete the following table

| Compounds | Lewis dot Structure | $A_xX_y$ Type | Electron Arrangement | Molecular Geometry | Bond Angle | Polarity |
|---|---|---|---|---|---|---|
| $SiH_4$ | | $AX_4$ | tetrahedral | tetrahedron | $109.5°$ | non-polar |
| $PCl_3$ | | $AX_3E$ | tetrahedral | Trigonal pyramid | $107°$ | polar |
| $H_2S$ | | $AX_2E_2$ | tetrahedral | bent | $105°$ | polar |
| $BeCl_2$ | | $AX_2$ | linear | linear | $180°$ | non-polar |
| $AlF_3$ | | $AX_3$ | trigonal planar | trigonal planar | $120°$ | non-polar |
| $SO_2$ | | $AX_2E$ | trigonal planar | bent | $120°$ | polar |
| $NO_3^-$ | | $AX_3$ | trigonal planar | trigonal planar | $120°$ | polar |

**Exercise 6.4 (p.75)**

1.    a)

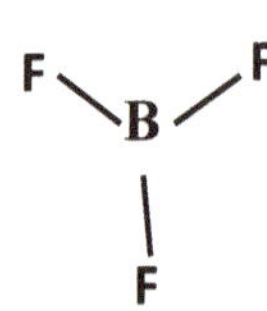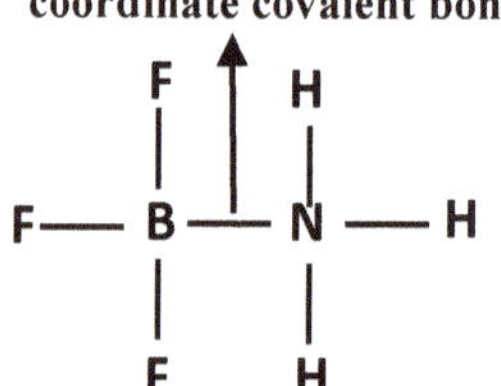


b)    **trigonal planar**        **trigonal pyramid**

c) **tetrahedral**        **d) coordinate covalent bond**

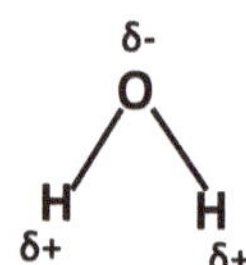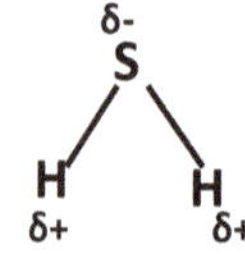

2.  Both have the same shape and polar bonds. The O-H bond in water is more polar than the S-H bond in hydrogen sulfide, since it has a greater electronegativity difference; O-H is 1.34 and S-H is 0.5. The intermolecular force between water molecules resulting from dipole-dipole attraction is thus greater than for hydrogen sulfide molecules. In addition, water molecules are attracted to each other by hydrogen bonds, which are absent in hydrogen sulfide molecules. These two combined forces are strong enough to allow water to be a liquid at room temperature. The relatively weaker intermolecular forces between the hydrogen sulfide molecules do not allow for them to be a liquid, but to exist as a gas.

3.  Since they have the same number of electrons, their London dispersion forces are the same. There is no electronegativity difference between the Br and Br atoms, so the molecules are non-polar. There is an electronegativity difference between the I and Cl atoms of 0.5, that makes the bond as well as the molecules polar. There is thus greater intermolecular force between I-Cl molecules than between $Br_2$ molecules, and for this reason I-Cl has a higher boiling point.

4.

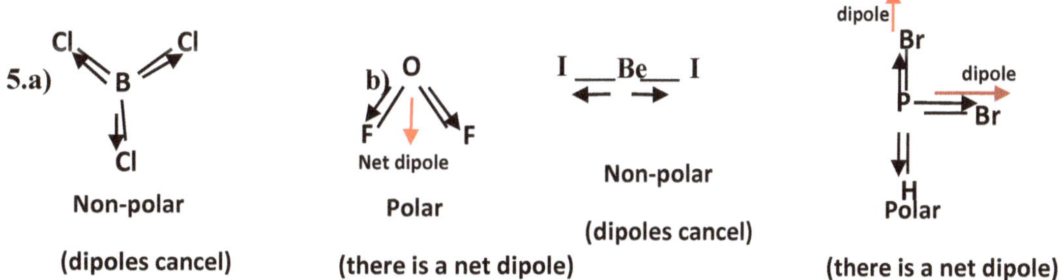


Both of these molecules are tetrahedrons. The $CH_4$ molecule is non-polar because of its symmetry even though the C-H bonds are polar. In the $CH_3Cl$, the C-Cl bond is much more polar than the three other C-H bonds. Because of this, the molecule is polar with a net dipole. The $CH_3Cl$ compound would have stronger intermolecular force than $CH_4$; $CH_3Cl$ having both L.D.F and dipole-dipole forces while $CH_4$ having only L.D.F. The boiling point of $CH_3Cl$ would be higher than that of $CH_4$.

5.a)

Cl — B — Cl, Cl

Non-polar

(dipoles cancel)

b)  O, F, F

Net dipole

Polar

(there is a net dipole)

I — Be — I

Non-polar

(dipoles cancel)

dipole

Br

P — Br  dipole

H

Polar

(there is a net dipole)

Exercise 6.5 (p.80)

1. (b) $CH_4$ (d) $C_2H_6$ (c) $C_3H_8$ (a) $C_4H_{10}$

2. (a) $Br_2$ (b) ICl

**Exercise 6.6 (p.81)**

1. a) 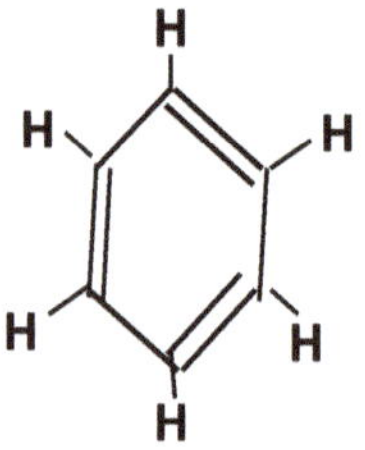

No effect

(symmetrical molecule)

b) Br—Br

No effect
(non-polar molecule)

c) 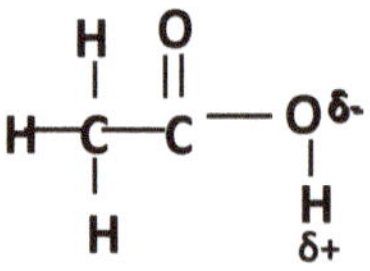

Attracted

(polar O-H bond)

d) 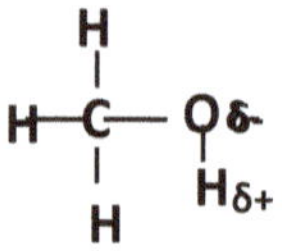

Attracted

(polar O-H bond)

# Chapter Review: Bonding (5-6) Solutions

## Matching

*Match each term in the table below with the correct statements that follow*

| | | | | |
|---|---|---|---|---|
| A | Octet | G | Instantaneous dipole |
| B | Electron cloud | H | Bonding capacity |
| C | Electric field | I | Polyatomic ion |
| D | Electronegativity | J | Coordinate covalent bond |
| E | Ionic bond | K | Heteronuclear |
| F | Polar | L | Dipole-dipole |

| | |
|---|---|
| 1. | This is produced by non-polar molecules as they polarize. |
| 2. | This is used to describe diatomic molecules with two different atoms. |
| 3. | This results when electrons are transferred from a metal to a non-metal. |
| 4. | The force of attraction that is present in polar molecules. |
| 5. | The number of covalent bonds an atom forms to complete its octet. |
| 6. | The volume of space around an atom or in a molecule where there is a probability of finding electrons. |
| 7. | An arrangement of eight electrons in the outer shell of an atom or ion. |
| 8. | A covalent bond where the two atoms have an electronegativity difference of more than 0.4. |
| 9. | A measure of the ability of an atom in a molecule to draw bonding electrons to itself. |
| 10. | A covalent bond that is formed one of the two bonded atoms contributes both electrons. |
| 11. | A number of non-metal atoms bonded together and having a charge. |
| 12. | This consists of electric field lines that are responsible for attracting opposite charges. |

## True or False

**Read each of the following statements and then decide if it is *True* or *False***

1. Atoms of elements combine with each other to achieve an inert gas electron configuration.

2. To gain an inert gas electron configuration, non-metals lose electrons.

3. When two or more nonmetals react, they form molecular compounds.

4. The bond formed when a metal reacts with a non-metal is ionic.

5. Covalent bonds are formed when electrons are transferred from one atom to another.

6. Covalent bonds are very weak.

7. A compound dissolves in water, so it must be ionic.

8. The smallest unit of an ionic compound is a molecule.

9. London Dispersion Forces are found in all molecular compounds.

10. The larger the molecule the weaker is the London Dispersion Force.

11. Ionic compounds dissolve in water to form electrolytes.

## Multiple Choice

**Choose the letter that best answers questions:**

1. Atoms of element attain an inert gas electron configuration by

| | | | | |
|---|---|---|---|---|
| a | sharing electrons | d | all of the above |
| b | gaining electrons | e | none of the above |
| c | losing electrons | | |

2. To attain an inert gas electron configuration, atoms of metallic elements

| | | | |
|---|---|---|---|
| a | share electrons | d | all of the above |
| b | gain electrons | e | none of the above |
| c | lose electrons | | |

3. The bonding capacity of nitrogen is

| | | | |
|---|---|---|---|
| a | 1 | d | 4 |
| b | 2 | e | 5 |
| c | 3 | | |

4. The valence attained by an ion depends on

| | | | |
|---|---|---|---|
| a | the number of electrons an atom gains | d | a and b |
| b | the number of electrons an atom loses | e | none of the above |
| c | the number of electrons an atom shares | | |

5. The formula for Iron (III) chloride is $FeCl_3$. In forming this compound,

| | | | |
|---|---|---|---|
| a | the iron atom loses 3 electrons | d | all of the above |
| b | each chlorine gains one electron | e | a and b |
| c | only the chlorine atoms the octet | | |

6. Covalent bonds are formed by reaction

| | | | |
|---|---|---|---|
| a | between metals and non-metals | d | between non-metals and metalloids only |
| b | between non-metals only | e | none of the above |
| c | between metals only | | |

7. A covalent bond is said to be negatively charged because

| | |
|---|---|
| a | in its formation one atom becomes negatively charged |
| b | a high concentration of negative cloud builds up between the two bonded atoms |
| c | a negative charge appears on the more electronegative atom |
| d | a and c |
| e | none of the above |

8. The number of **electron pairs** around the central nitrogen atom in $NH_3$ is?

| | | | |
|---|---|---|---|
| a | 4 | d | 1 |
| b | 3 | e | None of the above |
| c | 6 | | |

**9.**

In the $H_2CO_3$ molecule, the central carbon atom is surrounded by

| | | | | |
|---|---|---|---|---|
| a | 2 single bonds and 1 double bond | d | 2 double bonds |
| b | 3 single bonds | e | None of the above |
| c | 4 single bonds | | |

**10.**

Which of the following compounds is **not** ionic?

| | | | | |
|---|---|---|---|---|
| a | NaI | d | KI |
| b | HI | e | MgI$_2$ |
| c | LI | | |

**11.**

In which of the following compounds is the bond most polar?

| | | | | |
|---|---|---|---|---|
| a | HI | d | HBr |
| b | HCl | e | H$_2$S |
| c | HF | | |

**12.**

The bonds in the $CCl_4$ molecule are more polar than in the $NH_3$ molecule, yet $NH_3$ is polar but $CCl_4$ is not. The reason for this is due to the fact that

| | |
|---|---|
| a | $CCl_4$ molecule has no residual dipole but the $NH_3$ molecule has |
| b | the molecular geometry of $CCl_4$ is that of a tetrahedron while that of $NH_3$ is a trigonal pyramid |
| c | nitrogen is more electronegative than chlorine |
| d | a and b only |
| e | none of the above |

**13.**

The arrangement of the valence shell electron pairs in the $PH_3$ is?

| | | | | |
|---|---|---|---|---|
| a | trigonal planar | d | v-shaped |
| b | tetrahedral | e | None of the above |
| c | trigonal pyramidal | | |

**14.**

The shape of the $SO_2$ molecule is

| | | | | |
|---|---|---|---|---|
| a | linear | d. | tetrahedral |
| b | v-shaped | e. | none of the above |
| c | trigonal planar | | |

15. Which of the following compounds is polar?

| | | | | |
|---|---|---|---|---|
| *a* | $CO_{2(g)}$ | | *c.* | $PCl_3$ |
| *b* | $CH_4$ | | *d.* | $BF_3$ |
| *c* | $SiH_4$ | | | |

16. $H_2O$ is an example of a

| | | | | |
|---|---|---|---|---|
| *a* | $AX_4$ type of molecules | | *d.* | $AX_1E_3$ type of molecules |
| *b* | $AX_3E_1$ type of molecules | | *e.* | None of the above |
| *c* | $AX_2E_2$ type of molecules | | | |

17. Which of the following compounds has the highest boiling point?

| | | | | |
|---|---|---|---|---|
| *a* | $CH_4$ | | *d.* | $C_2H_6$ |
| *b* | $C_3H_8$ | | *e.* | $C_5H_{12}$ |
| *c* | $C_4H_{10}$ | | | |

18. London Dispersion Forces (L.D.S.) are associated with

| | | | | |
|---|---|---|---|---|
| *a* | electrostatic force | | *d.* | instantaneous dipoles |
| *b* | permanent dipoles | | *e.* | both c and d |
| *c* | induced dipoles | | | |

19. Carbon dioxide is 2.44 times as heavy as water; yet it is a gas and water is a liquid at STP. The reason for this is due to the fact that

| | |
|---|---|
| *a* | $CO_{2(g)}$ molecules are attracted by only weak London dispersion forces |
| *b* | $H_2O_{(l)}$ molecules are attracted only by stronger London dispersion forces |
| *c* | $H_2O_{(l)}$ molecules are attracted by both London dispersion forces and hydrogen bonds |
| *d* | $H_2O_{(l)}$ molecules are attracted by greater intermolecular forces than $CO_{2(g)}$ molecules |
| *e* | a, c and d |

20. HI is less polar than HCl, yet it has a higher boiling point than HCl. This is because

| | | | | |
|---|---|---|---|---|
| *a* | HI has stronger dipole-dipole forces | | *d.* | all of the above |
| *b* | HI has very strong L.D.F. | | *e.* | none of the above |
| *c* | HCl has no L.D.F. | | | |

# Chapter Review: Bonding (1-7) Solutions

## Matching:

    1.) G      2.) K      3.) E      4.) L      5.) H      6.) B      7.) A      8.) F

    9.) D      10.) J      11.) I      12.) C

## True/False

    1.) T      2.) F      3.) T      4.) T      5.) F      6.) F      7.) F      8.) F

    9.) T      10.) F      11.) T

## Multiple Choice

    1.) D      2.) C      3.) C      4.) D      5.) E      6.) B      7.) B      8.) A

    9.) A      10.) B      11.) C      12.) D      13.) B      14.) B      15.) C      16.) C

    17.) E      18.) E      19.) E      20.) B

# Chapter 7: Chemical Reactions

## Recognizing Chemical Reactions

**Exercise 7.1 (p.88)**

1. Evolution of $H_{2(g)}$
2. Formation of solid $Ag_2SO4_{(s)}$
3. Evolution of $O_{2(g)}$
4. Evolution of $O_{2(g)}$
5. Energy changes
6. Evolution of $HCl_{(g)}$ and $NH_{3(g)}$ or disappearance of $NH_4Cl_{(s)}$, a solid
7. Formation of solid $AgCl_{(s)}$
8. Formation of solid $CaCO_{3(s)}$
9. Formation of brown solid $Fe(OH)_{3(s)}$
10. Heat production

## Balancing Chemical Equations

**Exercise 7.2 (p.91)**

For the following word equations, write the chemical formulas of the reactants and products and then balance them:

a) sodium hydroxide + sulphuric acid $\rightarrow$ sodium sulphate + water
$$2\,NaOH_{(aq)} + H_2SO_{4(aq)} \rightarrow Na_2SO_{4(aq)} + 2\,H_2O_{(l)}$$
b) calcium hydroxide + nitric acid $\rightarrow$ calcium nitrate + water
$$Ca(OH)_{2(aq)} + 2\,HNO_{3(aq)} \rightarrow Ca(NO_3)_{2(aq)} + 2\,H_2O_{(l)}$$
c) hydrochloric acid + magnesium hydroxide $\rightarrow$ magnesium chloride + water
$$2\,HCl_{(aq)} + Mg(OH)_{2(aq)} \rightarrow MgCl_{2(aq)} + 2\,H_2O_{(l)}$$
d) sodium + water $\rightarrow$ sodium hydroxide + hydrogen (gas)
$$2\,Na_{(s)} + 2H_2O_{(l)} \rightarrow 2\,NaOH_{(aq)} + H_{2(g)}$$
e) potassium + oxygen $\rightarrow$ potassium oxide
$$4\,K_{(s)} + O_{2(g)} \rightarrow 2\,K_2O_{(s)}$$
f) zinc + sulphuric acid $\rightarrow$ zinc sulphate + hydrogen (gas)
$$Zn_{(s)} + H_2SO_{4(aq)} \rightarrow ZnSO_{4(aq)} + H_{2(g)}$$
g) magnesium + nitric acid $\rightarrow$ magnesium nitrate + hydrogen (gas)
$$Mg_{(s)} + 2HNO_{3(aq)} \rightarrow Mg(NO_3)_{2(aq)} + H_{2(g)}$$
h) copper + oxygen $\rightarrow$ copper (11) oxide
$$2\,Cu_{(s)} + O_{2(g)} \rightarrow 2\,CuO_{(s)}$$
i) sodium carbonate + hydrochloric acid $\rightarrow$ sodium chloride + water + carbon dioxide
$$Na_2CO_{3(aq)} + 2\,HCl_{(aq)} \rightarrow 2\,NaCl_{(aq)} + H_2O_{(l)} + CO_{2(g)}$$

**Exercise 7.3 (p.91)**

1. $4\,P_{(s)} + 3\,O_{2(g)} \rightarrow 2\,P_2O_{3(s)}$
2. $Zn_{(s)} + 2\,HCl_{(aq)} \rightarrow ZnCl_{2(aq)} + H_{2(g)}$
3. $2\,C_{(s)} + O_{2(g)} \rightarrow 2\,CO_{(g)}$
4. $Na_2CO_{3(aq)} + 2\,HCl_{(aq)} \rightarrow 2\,NaCl_{(aq)} + CO_{2(g)} + H_2O_{(l)}$
5. $2\,KNO_{3(s)} \rightarrow 2\,KNO_{2(s)} + O_{2(g)}$
6. $2\,KClO_{3(s)} \rightarrow 2\,KCl_{(s)} + 3\,O_{2(g)}$
7. $2\,H_2O_{2(l)} \rightarrow 2\,H_2O_{(l} + O_{2(g)}$
8. $2\,Al_{(s)} + 3\,CuCl_{2(aq)} \rightarrow 2\,AlCl_{3(aq)} + 3\,Cu_{(s)}$
9. $Ca(OH)_{2(aq)} + 2\,HNO_{3(aq)} \rightarrow Ca(NO_3)_{2(aq)} + 2\,H_2O_{(l)}$
10. $2\,AgNO_{3(aq)} + BaCl_{2(aq)} \rightarrow 2\,AgCl_{(s)} + Ba(NO_3)_{2(aq)}$

12. $Na_2CO_{3(aq)}$ + $CaCl_{2(aq)}$ → $CaCO_{3(s)}$ + 2 $NaCl_{(aq)}$
13. 2 $CO_{(g}$ + $O_{2(g)}$ → 2 $CO_{2(g)}$
14. $Fe_2O_{3(s)}$ + 3 $CO_{(g)}$ → 2 $Fe_{(s)}$ + 3 $CO_{2(g)}$
15. $Fe_2O_3$ + 6 $H_{2(g)}$ → 2 $Fe_{(s)}$ + 3 $H_2O_{(l)}$
18. 2 $Al(OH)_{3(s)}$ + 3 $H_2SO_{4(aq)}$ → $Al_2(SO_4)_{3(aq)}$ + 6 $H_2O_{(l)}$
19. $CO_{2(g)}$ + 2 $NH_{3(g)}$ → $CO(NH_2)_{2(aq)}$ + $H_2O_{(l)}$

## Types of Chemical Reactions

**Exercise 7.4 (p.93)**

**1**. For the following reactions, determine if they are synthesis or decomposition.
    (a)  3 $H_{2(g)}$ + $N_{2(g)}$ → 2 $NH_{3(g}$       Synthesis
    (b)  2 $HgO_{(s)}$ → 2 Hg + $O_{2(g)}$       Decomposition
    (c)  $H_2CO_{3(l)}$ → $H_2O_{(l)}$ + $CO_{2(g)}$       Decomposition
    (d)  $CaO_{(s)}$ + $H_2O_{(l)}$ → $Ca(OH)_{2(s)}$       Synthesis
    (e)  2 $NaHCO_{3(s)}$ → $Na_2CO_{3(s)}$ + $CO_{2(g)}$ + $H_2O_{(l)}$       Decomposition

**2**. The following are synthesis reactions. Complete the equations by first predicting the product, and then balancing them.
    (a) 2 $Ba_{(s)}$ + $O_{2(g)}$ → 2 $BaO_{(s)}$     (d) $Na_2O_{(s)}$ + $H_2O_{(l)}$ → 2 $NaOH_{(aq)}$
    (b) 2 $H_{2(g)}$ + $O_{2(g)}$ → 2 $H_2O_{(g)}$     (e) $SO_{3(g)}$ + $H_2O_{(l)}$ → $H_2SO_{4(aq)}$
    (c) 4 $Fe_{(s)}$ + 3 $O_{2(g)}$ → 2 $Fe_2O_{3(s)}$     (f) 2 $P_{(s)}$ + 3 $Cl_{2(g)}$ → 2 $PCl_{3(g)}$

**3**. The following are decomposition reactions. Complete the equations by first predicting the product, and then balancing them.
    (a) 2 $Ag_2O_{(s)}$ → 4 $Ag_{(s)}$ + $O_{2(g)}$     (d) $CaCO_{3(s)}$ → $CaO_{(s)}$ + $CO_{2(g)}$
    (b) 2 $H_2O_{2(l)}$ → 2 $H_2O_{(l)}$ + $O_{2(g)}$     (e) 2 $AlBr_{3(s)}$ → 2 $Al_{(s)}$ + 3 $Br_{2(g)}$
    (c) $K_2S_{(s)}$ → 2 $K_{(s)}$ + $S_{(s)}$     (f) 2 $CuO_{(s)}$ → 2 $Cu_{(s)}$ + $O_{2(g)}$

## Balancing Single Displacement Reactions

**Exercise 7.5 (p.100)**

**1**. Balance the following single displacement reactions equations:
    (a) $Cu_{(s)}$ + 2 $AgNO_{3(aq)}$ → $Cu(NO_3)_{2(aq)}$ + $Ag_{(s)}$
    (b) $Cl_{2(g)}$ + 2 $NaBr_{(aq)}$ → $NaCl_{(aq)}$ + $Br_{2(l)}$
    (c) 2 $Fe_{(s)}$ + 3 $SnCl_{2(aq)}$ → 2 $FeCl_{3(aq)}$ + 3 $Sn_{(s)}$
    (d) $Mg_{(s)}$ + 2 $HClO_{3(aq)}$ → $Mg(ClO_3)_{2(aq)}$ + $H_{2(g)}$
    (e) $Zn_{(s)}$ + 2 $HI_{(aq)}$ → $ZnI_{2(aq)}$ + $H_{2(g}$

**2**. The following are single displacement reactions. Complete the following equations by first predicting the products, and then balancing them.
    (b) $Br_{2(g)}$ + 2 $NaI_{(aq)}$ → $I_{2(s)}$ + 2 $NaBr_{(aq)}$
    (c) $Cu_{(s)}$ + 2 $AgNO_{3(aq)}$ → 2 $Ag_{(s)}$ + $Cu(NO_3)_{2(aq)}$
    (d) 4 $Al_{(s)}$ + 6 $H_2SO_{4(aq)}$ → 2 $Al_2(SO_4)_{3(aq)}$ + 3 $H_{2(g)}$
    (e) $Ca_{(s)}$ + 2 $H_2O_{(l)}$ → $Ca(OH)_{2(aq)}$ + $H_{2(g)}$
    (f) $F_2O_{3(s)}$ + 3 $CO_{(g)}$ → 2 $Fe_{(s)}$ + 3 $CO_{2(g)}$

**3.** Use the reactivity series to determine whether or not the following single reactions can happen. If a reaction is possible, complete the equation by first predicting the products, and then balance the equation. If a reaction cannot happen just write 'NR'.

$$
\begin{array}{llll}
\text{(a)} & Cu_{(s)} & + & SnCl_{2(aq)} & \rightarrow & NR \\
\text{(b)} & Cl_{2(g)} & + & 2\,NaI_{(aq)} & \rightarrow & I_{2(s)} & + & 2\,NaCl_{(aq)} \\
\text{(c)} & Ag_{(s)} & + & Cu(NO_3)_{2(aq)} & \rightarrow & NR \\
\text{(d)} & Zn_{(s)} & + & H_2O_{(l)} & \rightarrow & NR \\
\text{(e)} & Na_{(s)} & + & 2\,H_2O_{(l)} & \rightarrow & 2\,NaOH_{(aq)} & + & H_{2(g)} \\
\text{(f)} & Au_{(s)} & + & H_2SO_{4(aq)} & \rightarrow & NR \\
\text{(g)} & Br_{2(g)} & + & NaCl_{(aq)} & \rightarrow & NR \\
\text{(h)} & 2\,Al_{(s)} & + & 3\,SnSO_{4(aq)} & \rightarrow & Al_2(SO_4)_{3(aq)} & + & 3\,Sn_{(s)}
\end{array}
$$

## Exercise 7.6 (p.101)

$$
\begin{array}{lll}
\text{a)} & Mg_{(s)} & + & CuSO_{4(aq)} & \rightarrow & MgSO_{4(aq)} + Cu_{(s)} \\
\text{b)} & Cu_{(s)} & + & ZnSO_{4(aq)} & \rightarrow & NR \\
\text{c)} & Al_{(s)} & + & Mg(NO_3)_{2(aq)} & \rightarrow & NR \\
\text{d)} & Fe_{(s)} & + & CuSO_{4(aq)} & \rightarrow & FeSO_{4(aq)} + Cu_{(s)} \\
\text{e)} & 3\,Zn_{(s)} & + & 2\,Fe(NO_3)_{3(aq)} & \rightarrow & 3\,Zn(NO_3)_{2(aq)} + 2\,Fe_{(s)}
\end{array}
$$

# Double Displacement Reaction

## Exercise 7.8 (p.108)

For the following pairs of aqueous ionic compounds, complete the skeleton equations. If a double displacement reaction occurs, balance the equation. If no reaction occurs write NR.

$$
\begin{array}{llll}
\text{(a)} & 2\,NaOH_{(aq)} & + & H_2SO_{4(aq)} & \rightarrow & Na_2SO_{4(aq)} & + & 2\,H_2O_{(l)} + heat \\
\text{(b)} & 2\,KI_{(aq)} & + & Pb(NO_3)_{(aq)} & \rightarrow & 2\,KNO_{3(aq)} & + & PbI_{2(s)} \\
\text{(c)} & MgCl_{2(aq)} & + & NaNO_{3(aq)} & \rightarrow & NR \\
\text{(d)} & NH_4Cl_{(aq)} & + & KOH_{(aq)} & \rightarrow & KCl_{(aq)} + NH_{3(g)} + H_2O_{(l)} \\
\text{(e)} & H_2SO_{4(aq)} & + & 2\,CsOH_{(aq)} & \rightarrow & Cs_2SO_{4(aq)} & + & 2\,H_2O_{(l)} + heat \\
\text{(f)} & CaS_{(aq)} & + & 2\,HCl_{(aq)} & \rightarrow & CaCl_{2(sq)} & + & H_2S_{(g)} \\
\text{(g)} & ZnBr_{2(aq)} & + & HNO_{3(aq)} & \rightarrow & NR \\
\text{(h)} & K_2CO_{3(aq)} & + & 2\,HCl_{(aq)} & \rightarrow & 2\,KCl_{(aq)} + CO_{2(g)} + H_2O_{(l)}
\end{array}
$$

| Reagents | $HCl_{(aq)}$ $\frac{H^+ \mid Cl^-}{\text{ions}}$ | $K_2SO_{4(aq)}$ $\frac{K^+ \mid SO_4^{2-}}{\text{ions}}$ | $NaOH_{(aq)}$ $\frac{Na^+ \mid OH^-}{\text{ions}}$ | $ZnI_{2(aq)}$ $\frac{Zn^{2+} \mid I^-}{\text{ions}}$ | $Li_3PO_{4(aq)}$ $\frac{Li^+ \mid PO_4^{3-}}{\text{ions}}$ | $K_2S_{(aq)}$ $\frac{K^+ \mid S^{2-}}{\text{ions}}$ |
|---|---|---|---|---|---|---|
| $CuCl_{2(aq)}$ $\frac{Cu^{2+} \mid Cl^-}{\text{ions}}$ | NR | NR | NR | NR | $Cu_3(PO_4)_{2(s)} + LiCl_{(aq)}$ √ | √ $CuS_{(s)} + KCl_{(aq)}$ |
| $Ba(NO_3)_{2(aq)}$ $\frac{Ba^{2+} \mid NO_3^-}{\text{ions}}$ | NR | $BaSO_{4(s)} + KNO_{3(aq)}$ √ | NR | NR | $Ba_3(PO_4)_{2(s)} + LiNO_{3(aq)}$ √ | √ $BaS_{(s)} + KNO_{3(aq)}$ |
| $AgNO_{3(aq)}$ $\frac{Ag^+ \mid NO_3^-}{\text{ions}}$ | $AgCl_{(s)} + HNO_{3(aq)}$ √ | √ $Ag_2SO_{4(s)} + KNO_{3(aq)}$ | $AgOH_{(s)} + NaNO_{3(aq)}$ √ | √ $AgI_{(s)} + Zn(NO_3)_{2(aq)}$ | $Ag_3PO_{4(s)} + LiNO_{3(aq)}$ √ | √ $Ag_2S_{(s)} + KNO_{3(aq)}$ |
| $HNO_{3(aq)}$ $\frac{H^+ \mid NO_3^-}{\text{ions}}$ | NR | NR | $H_2O_{(l)} + NaNO_{3(aq)}$ + heat √ | NR | NR | $H_2S_{(g)} + KNO_{3(aq)}$ √ |
| $Al(NO_3)_{3(aq)}$ $\frac{Al^{3+} \mid NO_3^-}{\text{ions}}$ | NR | NR | $Al(OH)_{3(s)} + NaNO_{3(aq)}$ √ | NR | $Al_3PO_{4(s)} + LiNO_{3(aq)}$ √ | √ $Al_2S_{3(s)} + KNO_{3(aq)}$ |
| $Na_2CO_{3(aq)}$ $\frac{Na^+ \mid CO_3^{2-}}{\text{ions}}$ | $H_2O_{(l)} + CO_{2(g)} + NaCl_{(aq)}$ √ | NR | NR | $ZnCO_{3(s)} + NaI_{(aq)}$ √ | NR | NR |



## Combustion of Acetylene.

1.  Test tube 1.      **2.** Test tube 4.     **3.** Test tube 4.     **4.** Test tubes 1, 2 and 3.

5.  Test tube 4 produced the most energy (heat, light and sound) and there was no soot present.

    Test tubes 1, 2 and 3 produced less energy when compared to test tube 4. Also, soot was present to varying degrees in test tubes 1, 2 and 3.

6.  The more oxygen that was present, the greater was the degree of combustion.

7.  **Explanations of observations**

    **Test tube 1.** Oxygen was present only at the mouth of the test tube, so combustion took place only there and the flame was quickly extinguished. Since no oxygen was present in the test tube, no combustion was possible **inside** of it. The yellow flame was due to the presence of soot.

    **Test tube 2.** Oxygen was present throughout the test tube, so combustion took place throughout the test tube. However, the amount of oxygen present was insufficient for complete combustion. As a result of this, soot was present, and the flame was yellow. The other products are carbon dioxide and possibly some carbon monoxide.

**Test tube 3.** There was a greater oxygen to fuel ratio, so there was more combustion than in test tubes 1 and 2. However, the amount of oxygen was still inadequate for complete combustion as some trace of soot was present. The other products are carbon dioxide and possibly some carbon monoxide.

**Test tube 4.** The oxygen to fuel ratio was perfect so complete combustion took place. This is evident by the large amount of energy and blue flame produced, and the absence of any soot. Only carbon dioxide is possibly produced.

8.  Blue flame indicates complete combustion while yellow flame indicates incomplete combustion.

9.  Test tube 4.

**Making Connections**

10. It is important to tune up your vehicle for the following reasons:

    * A clogged-up air filter impedes the free passage of air for the combustion of the hydrocarbon fuel in the engine of automobiles. When this happens, there is incomplete combustion of the fuel which results in a lesser amount of energy being produced. Consequently, a larger amount of fuel is used to do the same amount of work. This represents a wastage of fuel and a greater risk for environmental pollution.

    *To ensure that the fuel is combusted completely, the wires that carry the current to the spark plugs to work efficiently must be changed periodically. Faulty wires would result in some spark plugs not firing, resulting in the fuel not being burnt in those cylinders. The unburnt fuel will be in the emitted gas mixture. For the same reason, periodically, spark plugs must be changed after they expire.

11. Burning in enclosed areas where there is lack of adequate supply of oxygen for complete combustion poses the risk of carbon monoxide poisoning.

12. For similar reasons given why it is necessary to tune up our vehicles, homeowners must have their furnaces regularly serviced and inspected. Faulty furnaces could result in incomplete combustion of fuel with the production of deadly carbon monoxide gas. The installation of carbon monoxide detectors in different parts of the home can quickly detect traces of carbon monoxide gas and potentially save lives.

**Ontario's Drive Clean**

13. (a) Vehicle A passed while vehicle B failed.

    (b) Check for faulty spark plugs, connecting wires, clogged-up air filters or possibly other problems related to the engine.

    (c) Vehicles with such types of emissions could make people suffer various types of cardiopulmonary diseases or may even worsen the states of those who are already suffering from them. These diseases include lung cancer, emphysema, asthma, silicosis and bronchitis.

15. Any emitted carbon monoxide can diffuse through the tiniest of spaces in the walls of houses. After an extended period, the concentration of carbon dioxide built up in the house can reach fatal levels.

* A small flaw in the numbering of the exercise in textbook.

> **Exercise 9.1 (p.116)**

a)  $NaBr_{(aq)} \rightarrow$ colorless       b) $FeCl_{3(aq)} \rightarrow$ brownish yellow   c) $AlCl_{3(aq)} \rightarrow$ colorless

d) $K_2Cr_2O_{7(aq)} \rightarrow$ orange     e) $CoCl_{2(aq)} \rightarrow$ pink   f) $CuSO_{4(aq)} \rightarrow$ blue

g) $Fe(NO_3)_2 \rightarrow$ light green    h) $Mg(NO_3)_2 \rightarrow$ colorless     i) $NiI_2 \rightarrow$ green

j) $CaBr_{2(aq)} \rightarrow$ colorless

## Exercise 8.1 (p.123)

| Compound | Formula | Compound | Formula |
|---|---|---|---|
| (a)  magnesium fluoride | $MgF_2$ | (k) aluminum nitride | AlN |
| (b) lead (II) oxide | PbO | (l)  tin (IV) bromide | $SnBr_4$ |
| (c)  potassium sulphide | $K_2S$ | (m) iron (II) chloride | $FeCl_2$ |
| (d) zinc iodide | $ZnI_2$ | (n) strontium nitride | $Sr_3N_2$ |
| (e) iron (III) sulphide | $Fe_2S_3$ | (o) sodium phosfide | $Na_3P$ |
| (f) calcium bromide | CaBr | (p) aluminum oxide | $Al_2O_3$ |
| (g) barium chloride | $BaCl_2$ | (q) cesium bromide | CsBr |
| (h) copper (II) oxide | CuO | (r) rubidium nitride | $Rb_3N$ |
| (i)  silver iodide | AgI | (s) gold (I) chloride | AuCl |
| (j) lead(IV) oxide | $PbO_2$ | (t) mercury(II) oxide | HgO |

## Exercise 8.2 (p.124)

(a)  $Li_2O$    lithium oxide  
(b)  CaO    calciumoxide  
(c)  $AlCl_3$    aluminum chloride  
(d)  MgS    magnesium sulphide  
(e)  $Na_2S$    sodium sulphide  
(f)  FeO    iron (II) oxide  
(g)  $CuCl_2$  copper (II) chloride  
(h)  $Na_3N$   sodium nitride  

(i) HgO    mercury (II) oxide  
(j) $Ag_2S$    silver sulphide  
(k) $PbCl_4$    lead (IV) chloride  
(l) $BaI_2$    barium iodide  
(m) $Cs_2O$    cesium oxide  
(n) $SnBr_4$    tin (IV) bromide  
(o) KF    potassium fluoride  
(p) $CoCl_2$    cobalt (II) chloride  

## Exercise 8.3 (p.125)

(a)  carbon tetrachloride $CCl_4$  
(b)  phosphorus trichloride $PCl_3$  
(c)  dinitrogen pentoxide $N_2O_5$  
(d)  sulphur hexachloride $SCl_6$  

(e)   nitrogen dioxide $NO_2$  
(f)   diphosphorus trioxide $P_2O_3$  
(g)   sulphur trioxide $SO_3$  
(h)   ammonia $NH_3$  

## Exercise 8.4 (p.125)

(a) $SBr_6$ sulphur hexabromide  
(b) $CS_2$ carbon disulphide  
(c) $SiO_2$ silicon dioxide  
(d) CO carbon monoxide  

(e)  $CF_4$ carbon tetrafluoride  
(f)  $P_2O_5$ diphosphorus pentoxide  
(g)  $HCl_{(g)}$ hydrogen chloride  
(h)  $N_2O_4$ dinitrogen tetroxide  

Write the IUPAC name for each of the following binary compounds.

(a) AuCl gold (I) chloride

(b) BaO barium oxide

(c) $PbI_2$ lead (II) iodide

(d) ZnO zinc oxide

(e) $HBr_{(g)}$ hydrogen bromide

(f) $SnO_2$ tin (IV) oxide

(g) $NCl_3$ nitrogen trichloride

(h) $K_3N$ potassium nitride

(i) CuO copper(II) oxide

(k) $Zn_3N_2$ zinc nitride

(l) $AuCl_3$ gold (III) chloride

(m) $Pb_3N_2$ lead (II) nitride

(n) ozone $O_{3(g)}$

(o) $Ag_3N$ silver nitride

(p) $Hg_2O$ mercury (1) oxide

(q) $N_2O$ dinitrogen monoxide

(r) $HI_{(g)}$ hydrogen iodide

(s) $OCl_2$ oxygen dichloride

# Chapter Review: Reactions (7-8) Solutions

## Matching

*Match each term in the table below with the correct statements that follow*

| | | | |
|---|---|---|---|
| A | $CH_{4(g)} + 2O_{2(g)} \rightarrow CO_{2(g)} + 2H_2O_{(l)} + energy$ | G | Decomposition |
| B | Catalyst | H | $Cu_{(s)} + 2HCl_{(aq)} \rightarrow CuCl_{2(aq)} + H_{2(g)}$ |
| C | Li, K Na, Ba and Ca | I | $SO_{3(g)} + H_2O_{(l)} \rightarrow H_2SO_{4(aq)}$ |
| D | $C_{(s)} + \dfrac{1}{2} O_{2(g)} \rightarrow CO_{(g)} + energy$ | J | $Mg_{(s)} + Cu^{2+}_{(aq)} \rightarrow Mg^{2+}_{(aq)} + Cu_{(s)}$ |
| E | $Zn_{(s)} + 2 HCl_{(aq)} \rightarrow ZnCl_{2(aq)} + H_{2(g)}$ | K | $4 Fe_{(s)} + 3O_{2(g)} \rightarrow 2 Fe_2O_{3(s)}$ |
| F | $HCl_{(aq)} + NaOH_{(aq)} \rightarrow NaCl_{(aq)} + H_2O_{(l)} + E$ | L | A change in temperature |

1. A reaction in which two or more substances combine to form a more complex compound.

2. A reaction in which a single complex compound breaks up into two or more simpler substances.

3. An example of a single displacement reaction.

4. An example of a double displacement reaction.

5. The net ionic equation of a single displacement reaction

6. One of the evidences of a chemical change.

7. An example of a combustion reaction.

8. An example of an incomplete combustion reaction.

9. An example of acid rain formation reaction.

10. This speeds up the rate of a chemical reaction without being used up.

11. Only these elements will displace hydrogen from water.

12. This reaction is not possible.

**True or False**

**Read each of the following statements and then decide if it is *True* or *False***

1. When aqueous solutions of two ionic compounds are mixed a reaction must happen.

2. Vinegar, when added to baking powder produces fizzing. A reaction thus takes place.

3. In a single displacement reaction, a single element reacts with an aqueous solution of a compound.

4. For a single displacement reaction to occur, the added element must be more reactive than one in the compound to which it added.

5. Any element above hydrogen in the reactivity series will displace hydrogen from dilute acids.

6. In decomposition reactions, less complex compounds are changed into two or more complex compounds.

7. In a double displacement reaction one of the products may be a precipitate.

8. Fluorine is the most reactive non-metal.

9. If aqueous solutions of silver nitrate and potassium iodide are mixed a precipitate of silver iodide is formed.

10. The reduction of iron (III) oxide by carbon dioxide is a double displacement reaction.

11. Carbon monoxide is produced as a result of incomplete combustion of carbon and hydrocarbons.

12. Neutralization reactions are examples of double displacement reactions in which there is no visible change.

13. Rusting of iron is a synthesis reaction.

14. Rusting of iron can be slowed down by coating it with a more reactive metal.

15. Acid rain is produced when non-metal oxides react with water in rain.

## Multiple Choice

**Choose the letter that best answers the question:**

1.

The indicator for a chemical change is

| | | | | |
|---|---|---|---|---|
| *a* | energy change | *d* | gas evolution |
| *b* | precipitate formation | *e* | all the above |
| *c* | colour change | | |

2.

$Mg_{(s)} + Zn(NO_3)_{2(aq)} \longrightarrow Zn_{(s)} + Mg(NO_3)_{2(aq)}$. The preceding equation is an example of a

| | | | |
|---|---|---|---|
| a | decomposition reaction | d. | combustion reaction |
| b | synthesis reaction | e. | single displacement reaction |
| c | double displacement reaction | | |

3.

In the presence of the catalyst manganese dioxide, hydrogen peroxide fizzes. This is an example of a

| | | | |
|---|---|---|---|
| a | synthesis reaction | d. | single displacement reaction |
| b | combustion reaction | e. | double displacement reaction |
| c | decomposition reaction | | |

4.

The following equation represents a single displacement reaction:

$$Cu_{(s)} + 2\,AgNO_{3(aq)} \rightarrow Cu(NO_3)_{2(aq)} + 2\,Ag_{(s)}$$

The observations made of this reaction over a time will be

| | |
|---|---|
| a | the formation of solid silver |
| b | the disappearance of the solid copper |
| c | the solution changes from colourless to light blue |
| d | the formation of a gas |
| e | a, b and c only |

5.

Chlorine gas (greenish-yellow in colour) is passed through a clear aqueous salt solution which then turned dark grey. The solution is most likely to be

| | | | |
|---|---|---|---|
| a | $NaCl_{(aq)}$ | d | $KF_{(aq)}$ |
| b | $NaBr_{(aq)}$ | e | none of the above |
| c | $KI_{(aq)}$ | | |

6.

A white solid compound dissolves in water, has a low melting point but does not conduct electricity. This compound is most likely to be

| | | | |
|---|---|---|---|
| a | polar and molecular | d | glucose |
| b | ionic | e | a and d |
| c | only molecular | | |

7.

Which of the following combinations will produce a single displacement reaction?

| | | |
|---|---|---|
| *a* | aqueous nitric acid and solid copper |
| *b* | aqueous sodium nitrate and solid aluminum |
| *c* | aqueous zinc iodide and solid lead |
| *d* | aqueous copper (II) chloride and solid iron |
| *e* | aqueous potassium nitrate and solid zinc |

8.

$$Fe_2O_{3(s)} + 2\ Al_{(s)} \longrightarrow Al_2O_{3(s)} + 2\ Fe_{(l)} + heat$$

The above reaction can be described as

| | | | | |
|---|---|---|---|---|
| *a* | double displacement | *d* | b and c |
| *b* | single displacement | *e* | decomposition |
| *c* | the Thermite reaction | | |

9.

Methane burns in oxygen according to the following equation:

$$CH_{4(g)} + 2\ O_{2(g)} \rightarrow CO_{2(g)} + 2\ H_2O_{(l)} + 890\ kJ/mol$$

In the absence of adequate amount of oxygen, the above reaction may

| | | | | |
|---|---|---|---|---|
| *a* | produce less energy | *d* | all of the above |
| *b* | also produce some carbon dioxide | *e* | none of the above |
| *c* | also produce some unburnt carbon | | |

10.

The following is an example of a double displacement reaction:

$$Na_2CO_{3(aq)} + 2\ HCl_{(aq)} \rightarrow 2\ NaCl_{(aq)} + H_2CO_{3(aq}\ (unstable\ and\ decomposes)$$

The evidence that this is a double displacement reaction will be

| | | | | |
|---|---|---|---|---|
| *a* | the formation of a precipitate | *d.* | a change in colour |
| *b* | the production of heat | *e.* | none of the above |
| *c* | the evolution of carbon dioxide gas | | |

11.

| The following reaction can be classified as a |||
|---|---|---|
| $Ca(OH)_{2(aq)} + 2\,HNO_{3(aq)} \rightarrow Ca(NO_3)_{2(aq)} + 2\,H_2O_{(l)}$ ||||
| *a* | neutralization reaction | *d.* | double displacement reaction |
| *b* | single displacement reaction | *e.* | a and d |
| *c* | decomposition reaction | | |

12.

| Which of the following mixtures of aqueous solutions will produce a double displacement reaction? ||
|---|---|
| *a* | nitric acid and potassium hydroxide solutions |
| *b* | sodium nitrate and potassium chloride solutions |
| *c* | zinc iodide and magnesium sulphate |
| *d* | potassium sulfate and iron (II) chloride solutions |
| *e* | sodium chloride and ammonium sulphate solutions |

13.

The following the following equation represents a double displacement reaction:

$$2\,KI_{(aq)} + Pb(NO_3)_{2(aq)} \rightarrow PbI_{2(s)} + 2\,KNO_{3(aq)}$$

The observation made of this reaction will be

| *a* | the formation of two aqueous solutions |
|---|---|
| *b* | the formation of yellow precipitate |
| *c* | the formation of a gas |
| *d* | the production of heat |
| *e* | none of the above |

14.

| Which of the following reactions represents acid rain formation? ||||
|---|---|---|---|
| *a* | $CO_{2(g)} + H_2O_{(l)} \rightarrow H_2CO_{3(aq)}$ | *d* | all of the above |
| *b* | $SO_{2(aq)} + H_2O_{(l)} \rightarrow H_2SO_{3(aq}$ | *e* | only a and b |
| *c* | $4\,NO_{2(g)} + 2\,H_2O_{(l)} \rightarrow 4\,HNO_{2(aq)}$ | | |

15.

| Which of the following may enter the atmosphere due to incomplete combustion of fuels or faulty catalytic converters in automobiles? ||||
|---|---|---|---|
| *a* | $CO_{(g)}$ | *d* | $C_{(s)}$ |
| *b* | $NO_{(g)}$ | *e* | all of the above |
| *c* | hydrocarbons | | |

16.

The addition of $AgNO_{3(aq)}$ to a sample of an aqueous salt solution produces a white precipitate that readily dissolves in dilute $NH_{3(aq)}$. The solution most likely contains

| | | | |
|---|---|---|---|
| a | $SO_4^{2-}{}_{(aq)}$ | d | $Br^-{}_{(aq)}$ |
| b | $CO_3^{2-}{}_{(aq)}$ | e | $I^-{}_{(aq)}$ |
| c | $Cl^-{}_{(aq)}$ | | |

17.

Zinc reacts with dilute hydrochloric acid according to the following equation:

$Zn_{(s)} + 2\,HCl_{(aq)} \rightarrow ZnCl_{2(aq)} + H_{2(g)}$ . The net ionic equation for this reaction is as follows:

$$Zn_{(s)} + 2H^+{}_{(aq)} \rightarrow Zn^{2+} + H_{2(g)}$$ . The half reactions for this are as follows:

$$* \quad Zn_{(s)} + energy \rightarrow Zn^{2+}{}_{(aq)} + 2e^-$$

$$* \quad 2\,H^+{}_{(aq)} + 2\,e^- \rightarrow H_{2(g)}$$

If the ionization energy of four metals: A, B, C and D are ranked: $A > B > C > D > E$.

Which of these will liberate hydrogen gas from dilute $HCl_{(aq)}$ at the fastest rate if they are all above hydrogen in the reactivity series?

| | | | |
|---|---|---|---|
| a | A | d | D |
| b | B | e | E |
| c | C | | |

18.

Carbonated sodas produce burping in the stomach when ingested. The reactions that take place in the stomach to produce this sensation can be described as:

| | | | |
|---|---|---|---|
| a | double displacement | d | single displacement reaction |
| b | decomposition | e | both a and b |
| c | synthesis | | |

19.

The following reaction: $2\,KClO_{3(aq)} \rightarrow 2\,KCl_{(aq)} + 3\,O_{2(g)}$ , can be classified as

| | | | |
|---|---|---|---|
| a | synthesis | d | single displacement |
| b | combustion | e | double displacement |
| c | decomposition | | |

20.

A solid compound produces a lilac flame when heated in Bunsen burner. When a sample of its aqueous solution reacts $BaCl_{2(aq)}$, a white precipitate is formed that re-dissolves in $HCl_{(aq)}$. The compound is most likely to be

| | | | |
|---|---|---|---|
| a | $Na_2CO_3$ | d | $K_2SO_3$ |
| b | $CuSO_4$ | e | $K_2SO_4$ |
| c | $KI$ | | |

21.

A sample of a blue aqueous salt solution reacts with $NH_{3(aq)}$ to produce a blue gelatinous precipitate that re-dissolves in excess $NH_{3(aq)}$ to form a deep blue solution. Another sample of it reacts with $BaCl_{2(aq)}$, to form a white precipitate that does no re-dissolve in excess $HCl_{(aq)}$. The salt solution is most likely to be

| | | | |
|---|---|---|---|
| a | $Na_2CO_3$ | d | $CuSO_3$ |
| b | KI | e | $K_2SO_4$ |
| c | $CuSO_4$ | | |

22.

The deep blue colour of the $[Cu(NH_3)_4(H_2O)_2]^{2+}$ complex is seen because the complex

| | |
|---|---|
| a | absorbs yellow and reflects deep blue |
| b | absorbs red and reflects deep blue |
| c | absorbs green and reflects deep blue |
| d | absorbs orange and reflects deep blue |
| e | none of the above |

23.

Ammonium chloride solution is mixed with a solution of o sodium hydroxide. If a colourless gas with a pungent odour that turns damp pink litmus paper bule, is produced, then the reaction can be described as

| | | | |
|---|---|---|---|
| a | double displacement | d. | single displacement reaction |
| b | decomposition | e. | none of the above |
| c | synthesis | | |

24.

The element iron form the following two compounds: $FeCl_2$ and $FeCl_3$. From this information, iron can be described as being

| | |
|---|---|
| a | multivalent |
| b | a transition element |
| c | univalent |
| d | a non-metal |
| e | a and b |

25.

In ionic compounds, the sum of the charges on the positive charges on the cations is equal to the sum of the negative charges on the anions. For the ion, $X^{3+}$, which of the following negative ion combination will produce a neutral compound with it.

| | |
|---|---|
| a | one $Cl^-$ ion with one $X^{3+}$ ion |
| b | three $Br^-$ ions with one $X^{3+}$ ion |
| c | three $O^{2-}$ ion with two $X^{3+}$ ions |
| d | One $N^{3-}$ ion $X^{3+}$ ion |
| e | b, c and d |

## Chapter Review: Reactions (7-8) Solutions

**Matching :**

1.) K        2.) G        3.) E        4.) F        5.) J        6.) L        7.) A        8.) D

9.) I        10.) B        11.) C        12.) H

**True/False**

1.) F        2.) T        3.) T        4.) T        5.) T        6.) T        7.) T        8.) T

9.) T        10.) T        11.) T        12.) T        13.) T        14.) T        15.) T

**Multiple Choice**

1.) E        2.) E        3.) C        4.) E        5.) C        6.) E        7.) D        8.) D        9.) D        10.) C        11.) E

12.) A        13.) B        14.) D        15.) E        16.) C        17.) E        18.) E        19.) C        20.) D        21.) C        22.) A

23.) A        24.) E        25.) E

# Unit 2: Quantities in Chemical Reactions



1. (a) true      (b) false      (c) false      (d) true
   (e) false      (f) true

2. (a) The number of protons
   (b) 6 protons in each of their atoms
   (c) 6 electrons in each of their atoms
   (d) $_6^{12}C$ has 6   $_6^{13}C$ has 7   $_6^{14}C$ has 8 neutrons each respectively.
   (e) The number of neutrons that they have.



| Atomic Notation | Number of Protons | Number of Neutrons | Number of Electrons | Mass Number |
|---|---|---|---|---|
| $_3^{7}Li$ | 3 | 4 | 3 | 7 |
| $_{11}^{23}Na$ | 11 | 12 | 11 | 23 |
| $_{20}^{40}Ca$ | 20 | 20 | 20 | 40 |
| $_{13}^{27}Al$ | 13 | 14 | 13 | 27 |
| $_{19}^{39}K$ | 19 | 20 | 19 | 39 |
| $_{15}^{31}P$ | 15 | 16 | 15 | 31 |

## The Mole Concept



1. (a) Na = 22.99g/mol   (b) Ca = 40.08 g/mol   (c) Cl = 35.45 g/mol   (d) Zn = 65.3g/mol

2. One mole or $6.02 \times 10^{23}$ atoms

3. Zn >, Ca >, Cl >Na

4. C

5. C-14

When solving problems in the mole concept involving pure substances (solids, liquids and gases), the following triangle is useful.

**The mole triangle**

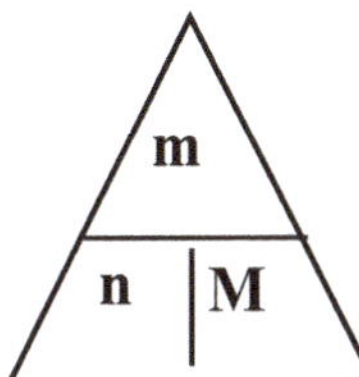

In this triangle:

n = number of moles

m = mass

M = molar mass

From this triangle the following three relationships can be obtained:

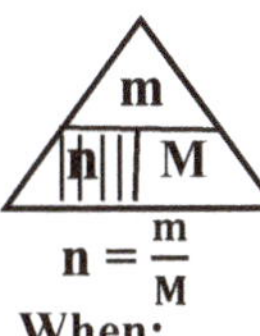

$$n = \frac{m}{M}$$

When:

m and M are given,
and n is required

$$m = n \times M$$

When:

n and M are given,
and m is required

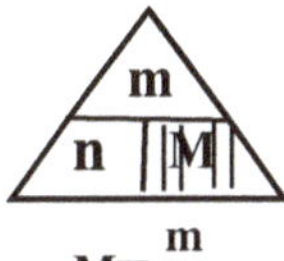

$$M = \frac{m}{n}$$

When:

m and n are given,
and M is required

**Exercise 9.4 (p.148)**

**Finding mass of elements, given their number of moles and molar masses.**

In solving these types of problems, the following triangle will be used:

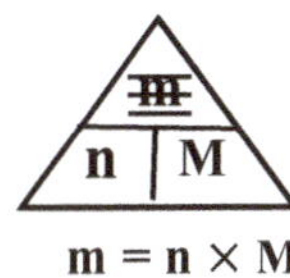

$$m = n \times M$$

(a) Find the mass of 5.00 mol of calcium.
Using the formula $m = n \times M$,

$$\text{we get } m = 5.00 \text{ mol} \times \frac{40.08 \text{ g}}{\text{mol}}$$

$$= 2.00 \times 10^2 \text{ g Ca}$$

(b) Find the mass of 2.50 mol of potassium

Using the formula $m = n \times M$,

$$\text{we get } m = 2.50 \text{ mol} \times \frac{39.10 \text{ g}}{\text{mol}}$$
$$= 97.8 \text{ g K}$$

(c) Find the mass of 3.00 mol of hydrogen. atoms.
Using the formula $m = n \times M$,

$$\text{we get } m = 3.00 \text{ mol} \times \frac{1.01 \text{g}}{\text{mol}}$$

$$= 3.03 \text{ g of H}$$

(d) Find the mass of 3.00 mol of chlorine atoms

Using the formula $m = n \times M$,

$$\text{we get } m = 4.00 \text{ mol} \times \frac{35.45 \text{ g}}{\text{mol}}$$
$$= 142 \text{ g Cl}$$

**Finding the number of moles of atoms, given their masses and molar masses**

In solving these types of problems, the following triangle is used.

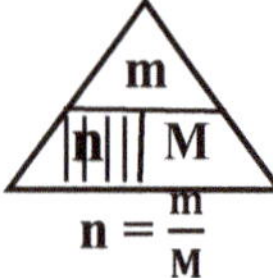

$$n = \frac{m}{M}$$

**(a)** Find the number of moles of atoms in 18.02 g Be.

Using the formula, $n = \frac{m}{M}$ we get $n = \dfrac{18.02\ g}{9.02\ g/mol} = \mathbf{2.00\ mol}$

**(b)** Find the number of moles of atoms in 72.93 g Mg.

Using the formula, $n = \frac{m}{M}$ we get $n = \dfrac{72.93\ g}{24.31\ g/mol} = \mathbf{3.000\ mol}$

**(c)** Find the number of moles of atoms in 60.02 g Ca.

Using the formula, $n = \frac{m}{M}$ we get $n = \dfrac{60.02\ g}{40.08\ g/mol} = \mathbf{1.497\ mol}$

**(d)** Find the number of moles of atoms in 68.97 g Na.

Using the formula, $n = \frac{m}{M}$ we get $n = \dfrac{68.97\ g}{22.99\ /mol} = \mathbf{3.000\ mol}$

## Calculating Number of Entities

To do calculations involving the number of entities such as atoms, molecules, ions or formula units, the following triangle is useful.

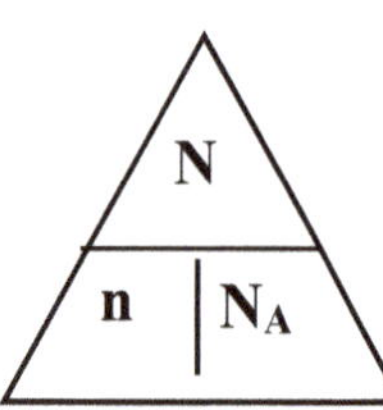

| **In this triangle:** |
| --- |
| $n$ = number of moles |
| $N$ = number of entities |
| $N_A$ = Avogadro's number |
| $= (6.022 \times 10^{23})$ |

From the triangle above, the following three relationships can be obtained:

 $\quad n = \dfrac{N}{N_A}$ $\qquad$  $\quad N = n \times N_A$ $\qquad$  $\quad N_A = \dfrac{N}{n}$

**When:**

N and $N_A$ are given, and
n is required

**When:**

n and $N_A$ are given, and
N is required

**When:**

N and n are given, and
$N_A$ is required

**(a)** Find the number of atoms in 6.00 moles of nitrogen atoms.

Using the formula, $N = n \times N_A$ we get $N = 6.00$ mol $\times 6.022 \times 10^{23}$ atoms/mol
$$= 3.61 \times 10^{24} \text{ atoms}$$

**(b)** Find the number of atoms in 2.50 moles of calcium atoms.

Using the formula, $N = n \times N_A$ we get $N = 2.50$ mol $\times 6.022 \times 10^{23}$ atoms/mol
$$= 1.51 \times 10^{24} \text{ atoms}$$

**(c)** Find the number of atoms in 4.00 moles of hydrogen atoms.

Using the formula, $N = n \times N_A$ we get $N = 4.00$ mol $\times 6.022 \times 10^{23}$ atoms/mol
$$= 2.41 \times 10^{24} \text{ atoms}$$

**(d)** Find the number of atoms in 3.00 moles of potassium atoms.

Using the formula, $N = n \times N_A$ we get $N = 3.00$ mol $\times 6.022 \times 10^{23}$ atoms/mol
$$= 1.81 \times 10^{24} \text{ atoms}$$

**(e)** Find the number of atoms in 5.00 moles of hydrogen atoms.

Using the formula, $N = n \times N_A$ we get $N = 5.00$ mol $\times 6.022 \times 10^{23}$ atoms/mol
$$= 3.01 \times 10^{24} \text{ atoms}$$

## Converting amounts in mass to number of moles

To do these types of calculations, *it is first necessary to convert amount in mass to amount in moles, then change that to number of entities.* The following two triangles are useful in these calculations:

$$n = \frac{m}{M} \qquad\qquad N = n \times N_A$$



**(a)** Find the number atoms in 4.00 g of oxygen.
    **Step 1.** Convert amount in mass to amount in moles.

Using the formula, $n = \frac{m}{M}$ we get $n = \dfrac{4.00 \text{g}}{16.00 \text{ g/mol}} = 0.250$ **mol**

    **Step 2.** Convert amount in moles to number of atoms.
Using the formula, $N = n \times N_A$ we get $N = 0.250$ mol $\times 6.022 \times 10^{23}$ atoms/mol
$$= 1.50 \times 6.022 \times 10^{23} \text{ atoms}$$

**(b)** Find the number atoms in 13.49 g of aluminum.
    **Step 1.** Convert amount in mass to amount in moles.

Using the formula, $n = \frac{m}{M}$ we get $n = \dfrac{13.49g}{26.98g/mol} = \mathbf{0.5000\ mol}$

**Step 2. v**
Using the formula, $N = n \times N_A$ we get $N = 0.5000$ mol $\times 6.022 \times 10^{23}$ atoms/mol
$$= \mathbf{3.011 \times 6.022 \times 10^{23}\,atoms}$$

**(c)** Find the number atoms in 92.9 g of phosphorus.
   **Step 1.** Convert amount in mass to amount in moles.

Using the formula, $n = \frac{m}{M}$ we get $n = \dfrac{9.9g}{30.97g/mol} = \mathbf{3.00\ mol}$

   **Step 2.** Convert amount in moles to number of atoms.
   Using the formula, $N = n \times N_A$ we get $N = 3.00$ mol $\times 6.022 \times 10^{23}$ atoms/mol
$$= \mathbf{1.81 \times 6.022 \times 10^{24}\,atoms}$$

**(d)** Find the number atoms in 126.78 g of copper.
   **Step 1.** Convert amount in mass to amount in moles.

Using the formula, $n = \frac{m}{M}$ we get $n = \dfrac{126.78g}{63.55g/mol} = \mathbf{1.995\ mol}$

   **Step 2.** Convert amount in moles to number of atoms.
   Using the formula, $N = n \times N_A$ we get $N = 1.995$ mol $\times 6.022 \times 10^{23}$ atoms/mol
$$= \mathbf{1.201 \times 6.022 \times 10^{24}\ atoms}$$

# Converting number of entities to amount in moles

To do these types of calculations, the following triangle is useful:

$$n = \frac{N}{N_A}$$



**(a)** Find the number of moles in $3.01 \times 10^{23}$ atoms of zinc.
   $3.01 \times 10^{23}$ atoms of zinc.
   Using the formula, we get, $n = \frac{N}{N_A}$ we get $3.01 \times 10^{23}$ atoms $\div 6.022 \times 10^{23}$ atoms/mol
$$= \mathbf{0.500\ mol\ Zn}$$

**(b)** Find the number of moles in $1.81 \times 10^{24}$ atoms of copper.
   Using the formula, we get, $n = \frac{N}{N_A}$ we get $1.81 \times 10^{24}$ atoms $\div 6.022 \times 10^{23}$ atoms/mol
$$= \mathbf{3.00\ mol\ Cu}$$

**(c)** Find the number of moles in $1.51 \times 10^{23}$ atoms of potassium.
   Using the formula, we get, $n = \frac{N}{N_A}$ we get $1.51 \times 10^{23}$ atoms $\div 6.022 \times 10^{23}$ atoms/mol
$$= \mathbf{0.250\ mol\ K}$$

**(d)** Find the number of moles in $6.02 \times 10^{22}$ atoms of lithium.

Using the formula, we get, $n = \frac{N}{N_A}$ we get $6.02 \times 10^{22}$ atoms $\div 6.022 \times 10^{23}$ atoms/mol
$$= \mathbf{0.100\ mol\ Li}$$

# Converting Number of Entities to Amounts in Mass

To do these types of calculations, *it is first necessary to convert the number of entities to amount in moles, then change that to amounts in mass.* The following two triangles are useful in these calculations:

$$n = \frac{N}{N_A} \qquad\qquad m = n \times M$$



**(a)** Convert $1.204 \times 10^{24}$ atoms of silver to amount in mass.

**Step 1.** Convert number of entities to number of moles.
Using the formula, $n = \dfrac{N}{N_A}$ we get $n = 1.204 \times 10^{24}$ atoms $\div\ 6.022 \times 10^{23}$ atoms/mol

$$= 2.000 \text{ mol}$$

**Step 2.** Convert number of moles to amount in mass.
Using the formula, $m = n \times M$ we get $m = 2.000$ mol $\times\ 107.87$ g/mol Ag

$$= 216 \text{ g Ag}$$

**(b)** Convert $6.02 \times 10^{22}$ atoms of zinc to amount in mass.
**Step 1.** Convert number of entities to number of moles.

Using the formula, $n = \dfrac{N}{N_A}$ we get $n = 6.02 \times 10^{22}$ atoms $\div\ 6.022 \times 10^{23}$ atoms/mol

$$= 0.100 \text{ mol}$$

**Step 2.** Convert number of moles to amount in mass.
Using the formula, $m = n \times M$ we get $m = 0.100$ mol $\times\ 65.39$ g/mol Zn

$$= 6.54 \text{ g Zn}$$

**(c)** Convert $3.01 \times 10^{23}$ atoms of lead to amount in mass.
**Step 1.** Convert number of entities to number of moles.
Using the formula, $n = \dfrac{N}{N_A}$ we get $n = 3.01 \times 10^{23}$ atoms $\div\ 6.022 \times 10^{23}$ atoms/mol Pb

$$= 0.500 \text{ mol}$$

**Step 2.** Convert number of moles to amount in mass.
Using the formula, $m = n \times M$ we get $m = 0.500$ mol $\times\ 207.20$ g/mol

$$= 20.7 \text{ g Pb}$$

**(d)** Convert $6.02 \times 10^{22}$ atoms of carbon to amount in mass.

**Step 1.** Convert number of entities to number of moles.
Using the formula, $n = \dfrac{N}{N_A}$ we get $n = 6.02 \times 10^{22}$ atoms $\div\ 6.022 \times 10^{23}$ atoms/mol

$$= 0.100 \text{ mol}$$

**Step 2.** Convert number of moles to amount in mass.
Using the formula, $m = n \times M$ we get $n = 0.100$ mol $\times\ 12.01$ g/mol

$$= 1.20 \text{ g C}$$

**Finding Molar Mass of Compounds**

Exercise 9.10 (p.155)

$CaCl_2$ -------- formula units

$CO_2$ --------- molecules

$Ca(NO_3)_2$ --- formula units

$PCl_3$ ---------- molecules

Exercise 9.11 (p.161)

**(a)** Find the molar mass of $Ca(OH)_2$.

$$Ca(OH)_2 \rightarrow 1\ Ca + 2\ O + 1\ H$$
$$1\ mol \rightarrow 1\ mol \quad 2\ mol \quad 1\ mol$$

$$\text{Molar mass} = 1\ mol \times \frac{40.08\ g}{mol} + 2\ mol \times \frac{16.00\ g}{mol} + 2\ mol \times \frac{1.01\ g}{mol}$$
$$= 40.08\ g + 32.00\ g + 2.02\ g$$

$$= \mathbf{74.1\ g/mol\ Ca(OH)_2}$$

**(b)** Find the molar mass of KOH.

$$KOH \rightarrow K + 1\ O + 1\ H$$
$$1\ mol \rightarrow 1\ mol \quad 1\ mol \quad 1\ mol$$

$$\text{Molar mass} = 1\ mol \times \frac{39.10\ g}{mol} + 1\ mol \times \frac{16.00\ g}{mol} + 1\ mol \times \frac{1.01\ g}{mol}$$
$$= 39.10\ g + 16.00\ g + 1.01\ g$$

$$= \mathbf{56.11\ g/mol\ KOH}$$

**(c)** Find the molar mass of $Fe_2O_3$.

$$Fe_2O_3 \rightarrow 2\ Fe + 3\ O$$
$$1\ mol \rightarrow 2\ mol + 3\ mol$$

$$\text{Molar mass} = 2\ mol \times \frac{55.85\ g}{mol} + 3\ mol \times \frac{16.00\ g}{mol}$$
$$= 111.7\ g + 48.00\ g$$
$$= \mathbf{159.7\ g/mol\ Fe_2O_3}$$

**(d)** Find the molar mass of $K_2Cr_2O_7$.

$$K_2Cr_2O_7 \rightarrow 2\ K + 2\ Cr + 7\ O$$
$$1\ mol \rightarrow 2\ mol \quad 2\ mol \quad 7\ mol$$

$$\text{Molar mass} = 2\ mol \times \frac{39.10\ g}{mol} + 2\ mol \times \frac{52.00\ g}{mol} + 7\ mol \times \frac{16.00\ g}{mol}$$
$$= 78.20\ g + 104.0\ g + 112.0\ g$$

$$= \mathbf{294.2\ g/mol\ K_2Cr_2O_7}$$

**(e)** Find the molar mass of $C_6H_{12}O_6$.

$$C_6H_{12}O_6 \rightarrow 6\ C + 12\ H + 6\ O$$
$$1\ mol \rightarrow 6\ mol \quad 12\ mol \quad 6\ mol$$

$$\text{Molar mass} = 6\ mol \times \frac{12.01\ g}{mol} + 12\ mol \times \frac{1.01\ g}{mol} + 6\ mol \times \frac{16.00\ g}{mol}$$
$$= 72.06\ g + 12.12\ g + 96.00\ g$$
$$= \mathbf{180.2\ g/mol\ C_6H_{12}O_6}$$

**(f)** Find the molar mass of $Ba(HCO_3)_2$.

$$Ba(HCO_3)_2 \rightarrow 1\ Ba \quad + \quad 2\ H \quad + \quad 2\ C \quad + \quad 6\ O$$

$$1\ mol \quad \rightarrow \quad 1\ mol \quad\quad 2\ mol \quad\quad 2\ mol \quad\quad 6\ mol$$

$$M = 1\ mol \times \frac{137.33\ g}{mol} + 2\ mol \times \frac{1.01\ g}{mol} + 2\ mol \times \frac{12.01\ g}{mol} + 6\ mol \times \frac{16.00\ g}{mol}$$

$$= 137.33\ g \quad + \quad 2.02\ g \quad + \quad 24.02\ g \quad + \quad 96.00\ g$$

$$= \textbf{259 g/mol } Ba(HCO_3)_2$$

**(g)** Find the molar mass of $Al_2(SO_4)_3$.

$$Al_2(SO_4)_3 \quad \rightarrow \quad 2\ Al \quad + \quad 3\ S \quad + \quad 12\ O$$

$$1\ mol \quad \rightarrow \quad 2\ mol \quad\quad 3\ mol \quad\quad 12\ mol$$

$$Molar\ mass = 2\ mol \times \frac{26.98\ g}{mol} + 3\ mol \times \frac{32.07\ g}{mol} + 12\ mol \times \frac{16.00\ g}{mol}$$

$$= 53.96\ g \quad + \quad 96.21\ g \quad + \quad 192\ g$$

$$= \textbf{342 g/mol } Al_2(SO_4)_3$$

**(h)** Find the molar mass of $N_2$.

$$N_2 \quad \rightarrow \quad 2\ N$$

$$1\ mol \quad \rightarrow \quad 2\ mol$$

$$Molar\ mass = 2\ mol \times \frac{14.01\ g}{mol}$$

$$= \textbf{28.02 g/mol } N_2$$

**(i)** Find the molar mass of $CO_2$.

$$CO_2 \quad \rightarrow \quad C \quad + \quad 2\ O$$

$$1\ mol \quad \rightarrow \quad 1\ mol \quad\quad 2\ mol$$

$$Molar\ mass = 1\ mol \times \frac{12.01\ g}{mol} + 2\ mol \times \frac{16.00\ g}{mol}$$

$$= 12.02\ g \quad + \quad 32.00\ g$$

$$= \textbf{44.02 g/mol } CO_2$$

**(j)** Find the molar mass of $Mg(NO_3)_2$.

$$Mg(NO_3)_2 \quad \rightarrow \quad 1\ Mg \quad + \quad 2\ N \quad + \quad 6\ O$$

$$1\ mol \quad \rightarrow \quad 1\ mol \quad\quad 2\ mol \quad\quad 6\ mol$$

$$Molar\ mass = 1\ mol \times \frac{24.31\ g}{mol} + 2\ mol \times \frac{14.01}{mol} + 6\ mol \times \frac{16.00\ g}{mol}$$

$$= 24.31\ g \quad + \quad 28.02\ g \quad + \quad 96.00\ g$$

$$= \textbf{148.3 g/mol } Mg(NO_3)_2$$

**(k)** Find the molar mass of $Ca(IO_3)_2$.

$$Ca(IO_3)_2 \quad \rightarrow \quad 1\ Ca \quad + \quad 2\ I \quad + \quad 6\ O$$

$$1\ mol \quad \rightarrow \quad 1\ mol \quad\quad 2\ mol \quad\quad 6\ mol$$

$$Molar\ mass = 1\ mol \times \frac{40.08\ g}{mol} + 2\ mol \times \frac{126.90\ g}{mol} + 6\ mol \times \frac{16.00\ g}{mol}$$

$$= 40.08\ g \quad + \quad 253.8\ g \quad + \quad 96.00\ g$$

$$= \textbf{389.9 g/mol } Ca(IO_3)_2$$

**(l)** Find the molar mass of $KMnO_4$.

$$KMnO_4 \quad \rightarrow \quad 1\ K \quad + \quad 1\ Mn \quad + \quad 4\ O$$

$$1\ mol \quad \rightarrow \quad 1\ mol \quad\quad 1\ mol \quad\quad 4\ mol$$

$$Molar\ mass = 1\ mol \times \frac{39.10\ g}{mol} + 1\ mol \times \frac{54.94\ g}{mol} + 4\ mol \times \frac{16.00\ g}{mol}$$

$$= 39.10\ g \quad + \quad 54.94\ g \quad + \quad 64.00\ g$$

$$= \textbf{158.0 g/mol } KMnO_4$$

**(m)** Find the molar mass of $CuSO_4.5H_2O$.

$$CuSO_4.5H_2O \rightarrow 1\ Cu + 1\ S + 9\ O + 10\ H \qquad 1\ mol \rightarrow$$
$$1\ mol \qquad 1\ mol \qquad 9\ mol \qquad 10\ mol$$

$$M = 1\ mol \times \frac{63.55\ g}{mol} + 1\ mol \times \frac{32.07\ g}{mol} + 9\ mol \times \frac{16.00\ g}{mol} + 10\ mol \times \frac{1.01\ g}{mol}$$
$$= 63.55\ g + 32.07\ g + 144.0\ g + 10.1\ g$$
$$= \mathbf{2.50 \times 10^2\ g/mol\ CuSO_4.5H_2O}$$

**(n)** Find the molar mass of $Ca(HCO_3)_2$.

$$Ca(HCO_3)_2 \rightarrow 1\ Ca + 2\ C + 6\ O + 2\ H$$
$$1\ mol \rightarrow 1\ mol \qquad 2\ mol \qquad 6\ mol \qquad 2\ mol$$

$$M = 1\ mol \times \frac{40.08\ g}{mol} + 2\ mol \times \frac{12.01\ g}{mol} + 6\ mol \times \frac{16.00\ g}{mol} + 2\ mol \times \frac{1.01\ g}{mol}$$
$$= 40.08\ g + 24.02\ g + 96.00\ g + 2.02\ g$$
$$= \mathbf{162\ g/mol\ Ca(HCO_3)_2}$$

**(o)** Find the molar mass of $H_2O_2$.

$$H_2O_2 \rightarrow 2\ H + 2\ O$$
$$1\ mol \rightarrow 2\ mol \qquad 2\ mol$$

$$\text{Molar mass} = 2\ mol \times \frac{1.01\ g}{mol} + 2\ mol \times \frac{16.00\ g}{mol}$$
$$= 2.02\ g + 32.00\ g$$
$$= \mathbf{34.02\ g/mol\ H_2O_2}$$

**(p)** Find the molar mass of $NaClO_3$.

$$NaClO_3 \rightarrow 1\ Na + 1\ Cl + 3\ O$$
$$1\ mol \rightarrow 1\ mol \qquad 1\ mol \qquad 3\ mol$$

$$\text{Molar mass} = 1\ mol \times \frac{22.99\ g}{mol} + 1\ mol \times \frac{35.45\ g}{mol} + 3\ mol \times \frac{16.00\ g}{mol}$$
$$= 22.99\ g + 35.45\ g + 48.00\ g$$
$$= \mathbf{106.4\ g/mol\ NaClO_3}$$

**(q)** Find the molar mass of $Ba(NO_3)_2$.

$$Ba(NO_3)_2 \rightarrow 1\ Ba + 2\ N + 6\ O$$
$$1\ mol \rightarrow 1\ mol \qquad 2\ mol \qquad 6\ mol$$

$$\text{Molar mass} = 1\ mol \times \frac{137.33\ g}{mol} + 2\ mol \times \frac{14.01\ g}{mol} + 6\ mol \times \frac{16.00\ g}{mol}$$
$$= 137.33\ g + 28.02\ g + 96.00\ g$$
$$= \mathbf{261.3\ g/mol\ Ba(NO_3)_2}$$

**(r)** Find the molar mass of $AgIO_3$.

$$AgIO_3 \rightarrow 1\ Ag + 1\ I + 3\ O$$
$$1\ mol \rightarrow 1\ mol \qquad 1\ mol \qquad 3\ mol$$

$$\text{Molar mass} = 1\ mol \times \frac{107.87\ g}{mol} + 1\ mol \times \frac{126.90\ g}{mol} + 3\ mol \times \frac{16.00}{mol}$$
$$= 107.87\ g + 126.90\ g + 48.00\ g$$
$$= \mathbf{282.8\ g/mol\ AgIO_3}$$

# Percentage Composition of Compounds

To do these types of problems, it is first necessary to find the molar masses of the compounds and then find the composition of each element in them.



Find the percentage composition, by mass, of the elements in the following compounds.

**(a)** $KNO_3$

$$KNO_3 \rightarrow 1\,K + 1\,N + 3\,O$$

$$1\,mol \rightarrow 1\,mol \quad 1\,mol \quad 3\,mol$$

$$\text{Molar mass} = 1\,mol \times \frac{39.10\,g}{mol\,K} + 1\,mol \times \frac{14.01g}{mol\,N} + 3\,mol \times \frac{16.00g}{mol\,O}$$

$$= 39.10\,g + 14.01\,g + 48.00\,g$$

$$= \mathbf{101.1\ g/mol}$$

Percentage composition of each element in the compound.

$$\% K = \frac{39.10\,g}{101.1\,g} \times 100\% \qquad \% N = \frac{14.01\,g}{101.1\,g} \times 100\% \qquad \% O = \frac{48.00\,g}{101.1\,g} \times 100\%$$

$$= \mathbf{38.67\ \%} \qquad\qquad = \mathbf{13.86\ \%} \qquad\qquad = \mathbf{47.48\ \%}$$

**(b)** $NH_4NO_3$

$$NH_4NO_3 \rightarrow 2\,N + 4\,H + 3\,O$$

$$1\,mol \rightarrow 2\,mol \quad 4\,mol \quad 3\,mol$$

$$\text{Molar mass} = 2\,mol \times \frac{14.01\,g}{mol\,N} + 4\,mol \times \frac{1.01g}{mol\,H} + 3\,mol \times \frac{16.00g}{mol\,O}$$

$$= 28.02\,g + 4.04\,g + 48.00\,g$$

$$= \mathbf{80.1\ g/mol}$$

Percentage composition of each element in the compound.

$$\% N = \frac{28.02\,g}{80.1\,g} \times 100\% \quad \% H = \frac{4.04\,g}{80.1\,g} \times 100\% \quad \% O = \frac{48.00\,g}{80.1\,g} \times 100\%$$

$$= \mathbf{35.0\ \%} \qquad\qquad = \mathbf{5.04\ \%} \qquad\qquad = \mathbf{59.91\ \%}$$

**(c)** $(NH_4)_2SO_4$

$$(NH_4)_2SO_4 \rightarrow 2\,N + 8\,H + 1\,S + 4\,O$$

$$1\,mol \rightarrow 2\,mol \quad 8\,mol \quad 1\,mol \quad 4\,mol$$

$$\text{Molar mass} = 2\,mol \times \frac{14.01\,g}{mol\,N} + 8\,mol \times \frac{1.01g}{mol\,H} + 1\,mol \times \frac{32.07g}{mol\,S} + 4\,mol \times \frac{16.00\,g}{mol\,O}$$

$$= 28.02\,g + 8.08\,g + 32.07\,g + 64.00\,g$$

$$= \mathbf{132\ g}$$

Percentage composition of each element in the compound.

$$\% N = \frac{28.02\,g}{132\,g} \times 100\% \quad \% H = \frac{8.08\,g}{132\,g} \times 100\% \quad \% S = \frac{32.07g}{132\,g} \times 100\% \quad \% O = \frac{64.00\,g}{132\,g} \times 100\%$$

$$= \mathbf{21.2\ \%} \qquad\quad = \mathbf{6.1\ \%} \qquad\quad = \mathbf{24.3\ \%} \qquad\quad = \mathbf{48.5\ \%}$$

**(d)** Comparing the percentage compositions of nitrogen in these compounds, it can be seen that, $NH_4NO_3$ **with 35.0 %,** will provide the greatest amount of nitrogen per gram.

**(e)** To solve these problems, it is necessary to first find the percentage composition of oxygen in the compounds.

i) NaOH $\rightarrow$ 1 Na + 1 O + 1 H

    1 mol $\rightarrow$ 1 mol    1 mol    1 mol

Find the molar mass of NaOH

$$\text{Molar mass} = 1 \text{ mol} \times \frac{22.99 \text{ g}}{\text{mol Na}} + 1 \text{ mol} \times \frac{16.00 \text{ g}}{\text{mol O}} + 1 \text{ mol} \times \frac{1.01 \text{ g}}{\text{mol H}}$$

$$22.99 \text{ g} + 16.00 \text{ g} + 1.01 \text{ g}$$

$$= \textbf{40.0 g/mol}$$

The percentage composition of oxygen in the compound will be:

$$\% \text{ O} = \frac{16.00 \text{ g}}{40.0 \text{ g}} \times 100\% = \textbf{40.0 \%.}$$

This means that in every 100.0 g of the compound, oxygen will comprise 40.0 g. Therefore, in 10.0 g of NaOH, the mass of oxygen will be:

$$\frac{40.0 \text{ g}}{100.0 \text{ g}} \times 10 \text{ g} = \textbf{4.00 g O}$$

ii) $H_2SO_4$ $\rightarrow$ 2 H + 1 S + 4 O

    1 mol $\rightarrow$ 2 mol + 1 mol + 4 mol

Find the molar mass of $H_2SO_4$.

$$\text{Molar mass} = 2 \text{ mol} \times \frac{1.01 \text{ g}}{\text{mol H}} + 1 \text{ mol} \times \frac{32.07 \text{ g}}{\text{mol S}} + 4 \text{ mol} \times \frac{16.00 \text{ g}}{\text{mol O}}$$

$$= 2.02 \text{ g} + 32.07 \text{ g} + 64.00 \text{ g}$$

$$= \textbf{98.09 g/mol}$$

The percentage composition of oxygen in the compound will be:

$$\% \text{ O} = \frac{64.00 \text{ g}}{98.09 \text{ g}} \times 100\% = \textbf{65.24 \%.}$$

This means that in every 100.0 g of the compound, oxygen will comprise 65.24 g. Therefore, in 49.0 g of $H_2SO_4$, oxygen will be:

$$\frac{65.25 \text{ g}}{100 \text{ g}} \times 49.0 \text{ g} = \textbf{32.0 g O}$$

# Finding Empirical and Molecular Formulas of Compounds



## *Problem:*

1. Glucose has the following percentage composition by mass: 39.99% carbon, 6.73% hydrogen, and 53.28% oxygen. If the molar mass of glucose is 180.18 g/mol, what is its molecular formula? How many moles of glucose are required if each of 200 athletes is to be given a glucose tablet of having a mass of 10.0 g?

**Step 1**. Convert percentage composition to amount in mass (g).

| Elements: | C | : | H | : | O |
|---|---|---|---|---|---|
| Mass: | 39.99 g | | 6.73 g | | 53.28 g |

**Step 2**. Find the number of moles of each element present in the compound.

$$\frac{39.99 \text{ g}}{12.01\frac{g}{mol}C} \qquad \frac{6.73 \text{ g}}{1.01\frac{g}{mol}H} \qquad \frac{53.28 \text{ g}}{16.00\frac{g}{mol}O}$$

$$\textbf{3.33 mol} \quad : \quad \textbf{6.66 mol} \quad : \quad \textbf{3.33 mol}$$

**Step 3.** To get rid of decimals, divide all the numbers by the smallest one.

$$\frac{3.33 \text{ mol}}{3.33} \qquad \frac{6.66 \text{ mol}}{3.33} \qquad \frac{3.33 \text{ mol}}{3.33}$$

$$1 \text{ mol} \quad : \quad 2 \text{ mol} \quad : \quad 1 \text{ mol}$$

Making each of these the subscript of the elements in the compound we get for its empirical formula, $CH_2O$.

**Step 4.** Find the molar mass for the empirical formula, $CH_2O$.

$$M = 1 \times 12.01 \text{ g/mol C} + 2 \times 2.02 \text{ g/mol H} + 1 \times 16.00 \text{ g/mol O} = \textbf{\textit{30.03 g/mol}}$$

**Step 5.** Divide the molar mass of the compound by the molar mass of its empirical formula.

$$= \frac{180.18 \text{ g/mol}}{30.03 \text{ g /mol}} = 6$$

**Step 6.** Multiply the subscript of each of the elements in the molecular formula by **6**.

$$\text{Molecular formula} = C_{1\times 6}\times H_{2\times 6}O_{1\times 6} = C_6H_{12}O_6$$

Mass of glucose needed for 200 athletes $= 200 \times 10.0 \text{ g} = \textbf{3.00} \times^2 \textbf{g}.$

Number of moles of glucose $= 3.00 \times^2 \text{ g} \div 180.18 \text{ g/mol} = \textbf{1.67 mol}$

## *Problem:*

2. The hydrocarbon, butane, has the following percentage composition by mass: 82.63% carbon, and 17.37% hydrogen. If the molar mass of butane is 58.14 g/mol, what is its molecular formula?

**Step 1.** Convert percentage composition to amount in mass (g).

| Elements: | C | : | H |
|---|---|---|---|
| Mass: | 82.63 g | | 17.37 g |

**Step 2.** Find the number of moles of each element present in the compound.

$$\frac{82.63 \text{ g}}{12.01\frac{g}{mol}C} \qquad \frac{17.37 \text{ g}}{1.01\frac{g}{mol}H}$$

$$\textbf{6.88 mol} \quad : \quad \textbf{17.20 mol}$$

**Step 3.** To get rid of decimals, divide all the numbers by the smallest one; **6.88.**

$$\frac{6.88\ mol}{6.88} \qquad\qquad \frac{17.20\ mol}{6.88}$$

$$\textbf{1 mol} \qquad : \qquad \textbf{2.50 mol}$$

**Step 4.** Get rid of the ensuing decimal, multiply all values by **2.** Doing this we get

$$\textbf{2} \qquad : \qquad \textbf{5}$$

Making each of these the subscript of the elements in the compound, we get for its empirical formula, $\textbf{C}_2\textbf{H}_5\textbf{.}$

**Step 5.** Find the molar mass for the empirical formula, $C_2H_5$.

$$M\ =\ 2 \times 12.01\ \text{g/mol C}\ +\ 5 \times 1.01\ \text{g/mol H}\ \ =\ \textit{29.06 g/mol}$$

**Step 6.** Divide the molar mass of the compound by the molar mass of its empirical formula.

$$=\ \frac{58.14\ \text{g/mol}}{29.06\ \text{g /mol}}\ =\textbf{2}$$

**Step 7.** Multiply the subscript of each of the elements in the molecular formula by **2.**

$$\text{Molecular formula} = C_{2\times2}H_{5\times2}\ =\ \textbf{C}_4\textbf{H}_{10}$$

## *Problem:*

**3.** Caffeine (trimethylxanthine), the stimulant found in tea and coffee, has the following percentage composition by mass: 49.47% carbon, 5.20% hydrogen, 28.85% nitrogen, and 16.48% oxygen. What is its empirical formula?

**Step 1.** Convert percentage composition to amount in mass (g).

| Elements: | C | : | H | : | N | : | O |
|---|---|---|---|---|---|---|---|
| Mass: | 49.47 g | | 5.20 g | | 28.85 | | 16.48 g |

**Step 2.** Find the number of moles of each element present in the compound.

$$\frac{49.47\ \text{g}}{12.01\frac{g}{mol}\text{C}} \qquad \frac{5.20\ \text{g}}{1.01\frac{g}{mol}\text{H}} \qquad \frac{28.85\ \text{g}}{\frac{14.01g}{mol}\text{N}} \qquad \frac{16.48\ \text{g}}{16.00\frac{g}{mol}\text{O}}$$

$$\textbf{4.12} \quad : \quad \textbf{5.15} \quad : \quad \textbf{2.06} \quad : \quad \textbf{1.03}$$

**Step 3.** To get rid of decimals, divide all the numbers by the smallest one, **1.03.**

$$\frac{4.12}{1.03} \qquad \frac{5.14}{1.03} \qquad \frac{2.06}{1.03} \qquad \frac{1.03}{1.03}$$

$$\textbf{4} \quad : \quad \textbf{5} \quad : \quad \textbf{2} \quad : \quad \textbf{1}$$

Making each of these the subscript of the elements in the compound, we get for its empirical formula, $\textbf{C}_4\textbf{H}_5\textbf{N}_2\textbf{O}\textbf{.}$

## *Problem:*

**4.** Vitamin C, a deficiency of which causes the disease scurvy, has the following percentage composition by mass: 40.91% carbon, 4.59% hydrogen, and 54.50% oxygen. If the molar mass of vitamin C is 176.14 g/mol, what is its molecular formula? If you take a 60.0 mg dose of this vitamin daily, how long will it take for you to consume one mole?

**Step 1.** Convert percentage composition to amount in mass (g).

| Elements: | C | : | H | : | O |
|---|---|---|---|---|---|
| Mass: | 40.91 g | | 4.59 g | | 54.50 g |

**Step 2.** Find the number of moles of each element present in the compound.

$$\frac{40.91 \text{ g}}{12.01\frac{g}{mol}C} \qquad \frac{4.59 \text{ g}}{1.01\frac{g}{mol}H} \qquad \frac{54.50 \text{ g}}{16.00\frac{g}{mol}O}$$

$$\textbf{3.410 mol} \quad : \quad \textbf{4.54 mol} \quad : \quad \textbf{3.410 mol}$$

**Step 3.** To get rid of decimals, divide all the numbers by the smallest one.

$$\frac{3.410 \text{ mol}}{3.410} \qquad \frac{4.54 \text{ mol}}{3.410} \qquad \frac{3.410 \text{ mol}}{3410}$$

$$\textbf{1 mol} \quad : \quad \textbf{1.33 mol} \quad : \quad \textbf{1 mol}$$

**Step 4**. Get rid of the ensuing decimal, multiply all values by **3**. Doing this we get

$$\textbf{3} \qquad : \qquad \textbf{4} \qquad : \qquad \textbf{3}$$

Making each of these the subscript of the elements in the compound we get for its empirical formula, $C_3H_4O_3$.

**Step 5.** Find the molar mass for the empirical formula, $C_3H_4O_3$.

$$\text{M} \quad = 3 \times 12.01 \text{ g/mol C} \ + \ 4 \times 1.01 \text{ g/mol H} \ + \ 3 \times 16.00 \text{ g/mol O} \ = \textbf{\textit{88.1 g/mol}}$$

**Step 6**. Divide the molar mass of the compound by the molar mass of its empirical formula.

$$= \ \frac{176.14 \text{ g/mol}}{88.1 \text{ g /mol}} = \textbf{2}$$

**Step 7.** Multiply each of the subscripts in the empirical formula by **2** to get the molecular formula.

$$C_{3\times2}H_{4\times2}O_{3\times2} \ = \ \textbf{C}_\textbf{6}\textbf{H}_\textbf{8}\textbf{O}_\textbf{6}$$

The molar mass = 6 × 12.01 g/mol C + 8 × 1.01 g/mol H + 6 × 16.00 g/mol O = **176.14 g/mol**

The mass of one dose = 60.0 mg ÷ 100 g/mg  = **0.0600 g**

Since this amount is consumed in one day, the time it will take to consume one mole will be,

$$\textbf{176.14 g} \div \ 0.060 \text{ g/day} = \textbf{1936 days}$$

***Problem:***

**5.** If the molecular formula for vitamin D which is responsible for calcium metabolism is $C_{56}H_{88}O_2$ find its percentage composition by mass of the elements present.

(a) $C_{56}H_{88}O_2 \quad \rightarrow \quad$ C $\quad + \quad$ H $\quad + \quad$ 2 O

$\qquad$ 1 mol $\quad \rightarrow \quad$ 56 mol $\quad + \quad$ 88 mol $\quad + \quad$ 2 mol

$$\text{Molar mass} \quad = 56 \text{ mol} \times \frac{12.01 \text{ g}}{mol \text{ C}} \ + \ 88 \text{ mol} \times \frac{1.01 g}{mol \text{ H}} \ + \ 2 \text{ mol} \times \frac{16.00 g}{mol \text{ O}}$$

$$= \qquad 672.6 \text{ g} \qquad + \qquad 88.8 \text{ g} \qquad + \qquad 32.00 \text{ g}$$

$$= \textbf{793.4 g/mol}$$

Percentage composition of each element in the compound.

$$\% \text{ C} = \frac{672.6 \text{ g}}{793.4 \text{ g}} \times 100\% \ : \ \% \text{ H} = \frac{88.8 \text{ g}}{793.4 \text{ g}} \times 100\% \ : \ \% \text{ O} = \frac{32.00 \text{ g}}{793.4 \text{ g}} \times 100\%$$

$$= \textbf{84.79 \%} \qquad\qquad = \textbf{11.2 \%} \qquad\qquad = \textbf{4.03 \%}$$

## Problem:

6.  Folic acid, (one of the 8 B vitamins), is important for good health and for the reduction of spina bifida in unborn children. If it has the molecular formula, $C_{19}H_{19}N_7O_6$, find its percentage composition by mass of the elements present. If the daily requirement for the male adult is 400 micrograms, how many vitamin B tablets can be manufactured from 10 moles of folic acid?

(a) Find the molar mass of $C_{19}H_{19}N_7O_6$.

$$C_{19}H_{19}N_7O_6 \rightarrow 19\,C + 19\,H + 7\,N + 6\,O$$

$$1\ mol \rightarrow 19\ mol + 19\ mol + 7\ mol + 6\ mol$$

$$M = 19\ mol \times \frac{12.01\ g}{mol\ C} + 19\ mol \times \frac{1.01\ g}{mol\ H} + 7\ mol \times \frac{14.01\ g}{mol\ N} + 6\ mol \times \frac{16.00\ g}{mol\ O}$$

$$= 228.2\ g + 19.19\ g + 98.07\ g + 96.00\ g$$

$$= \mathbf{441.4\ g/mol}$$

(b) Percentage composition:

$$C = \frac{228.2\ g}{441.4g} \times 100\% \ : \ H = \frac{19.19\ g}{441.4\ g} \times 100\% : N = \frac{98.07\ g}{441.4\ g} \times 100\% \ : O = \frac{96.00\ g}{441.4\ g} \times 100\%$$

$$= \mathbf{51.69\ \%} \qquad = \mathbf{4.347\ \%} \qquad = \mathbf{22.22\ \%} \qquad = \mathbf{21.74\ \%}$$

(c) Find the mass of 10 moles of $C_{19}H_{19}N_7O_6$

Using the formula, $\mathbf{m = n \times M}$, we get

Mass = 10 mol × 441.4 g/mol  = **4414 g**

The mass of each tablet in grams = 400 microgram = 0.400 g/tablet

$$= \mathbf{4 \times 10^{-4}\ g}$$

The number of tablets that can be manufactured = 4414 g ÷ 0.400 g/tablet

$$= \mathbf{11035\ tablets}$$

## Problem:

7. Lactic acid is the substance that causes muscle fatigue in athletes. Its percentage composition by mass is: 40.03% carbon, 6.72% hydrogen, and 53.25% oxygen. If its molar mass is **90.15 g/mol,** determine its molecular formula.

**Step 1.** Convert percentage composition to amount in mass (g).

| Elements: | C | : | H | : | O |
|---|---|---|---|---|---|
| Mass: | 40.03 g | | 6.72 g | | 53.25 g |

**Step 2**. Find the number of moles of each element present in the compound.

$$\frac{40.03\ g}{12.01\frac{g}{mol}\ C} \qquad \frac{6.72\ g}{1.01\frac{g}{mol}\ H} \qquad \frac{53.25\ g}{16.00\frac{g}{mol}\ O}$$

$$\mathbf{3.330\ mol} \quad : \quad \mathbf{6.65\ mol} \quad : \quad \mathbf{3.330\ mol}$$

**Step 3**. To get rid of decimals, divide all the numbers by the smallest one.

$$\frac{3.330\ mol}{3.330} \qquad \frac{6.65\ mol}{3.330} \qquad \frac{3.330\ mol}{3.330}$$

$$\mathbf{1\ mol} \quad : \quad \mathbf{2\ mol} \quad : \quad \mathbf{1\ mol}$$

Making each of these the subscript of the elements in the compound we get for its empirical formula, $C_1H_2O_1$.

**Step 4.** Find the molar mass for the empirical formula, $C_1H_2O_1$.

M = 1×12.01 g/mol C  + 2 × 1.01 g/mol H  + 1×16.00 g/mol O  = *30.03 g/mol*

**Step 5.** Divide the molar mass of the compound by the molar mass of its empirical formula.

90.15 g/mol ÷ 30.03 g/mol  = **3**

**Step 6.** Multiply the subscript of each of the elements in the empirical formula by **3.**

Molecular formula = $C_{1×3}H_{2×3}O_{1×3}$ = **$C_3H_6O_3$**

## *Problem:*

**8.** If the artificial sweetener, aspartame, has the following percentage composition by mass: 57.12% carbon, 6.18 % hydrogen, 9.52 % nitrogen, and 27.0 % oxygen determine its empirical formula. How many 0.50 g packages can be made from 10 moles of aspartame?

**Step 1.** Convert percentage composition to amount in mass (g).

| Elements: | C | : | H | : | N | : | O |
|---|---|---|---|---|---|---|---|
| Mass: | 57.12 g | | 6.18 g | | 9.52 g | | 27 g |

**Step 2.** Find the number of moles of each element present in the compound.

$$\frac{57.12\ g}{12.01\frac{g}{mol}\ C} \qquad \frac{6.18\ g}{1.01\frac{g}{mol}\ H} \qquad \frac{9.52\ g}{14.01\frac{g}{mol}\ N} \qquad \frac{27.00\ g}{16.00\frac{g}{mol}\ O}$$

**4.760 mol** : **6.12 mol** : **0.679 mol** : **1.680 mol**

**Step 3.** To get rid of decimals, divide all the numbers by the smallest one, **0.679 mol.**

$$\frac{4.760\ mol}{0.679} \qquad \frac{6.12\ mol}{0.679} \qquad \frac{0.679\ mol}{0.679} \qquad \frac{1.680\ mol}{0.679}$$

**7.01 mol** : **9.01 mol** : **1.00 mol** : **2.48 mol**

**Step 4.** Get rid of the ensuing decimal, multiply all values by **2**. Doing this we get

**14** : **18** : **2** : **5**

Making these the subscripts in for the elements in the compound, we get for the molecular formula; **$C_{14}H_{18}N_2O_5$.**

M = 14 × 12.01g/mol C + 18 × 1.01 g/mol H + 2 × 14.01 g/mol N + 5 × 16.00 g/mol O

168.14 g + 18.18 g + 28.02 g + 80.00 g = *294.3 g/mol*

The mass of 12 moles of aspertame is found by using the formula:

**m = n × M** = 10 mol × 294.3 g/mol  = **2943 g**

Mass of one pack of aspartame = **0.50 g/pack**

m = n × M

The number of packs that can be made from 2943 g = **2943 g ÷ 0.50 g/pack**

= **5886 packs**

## *Problem:*

**9.** Morphine is one of the most widely used drugs to kill pain. Its percentage composition by mass is: 71.54% C, 6.72% H, 4.91% N, and 16.82 % O. If its molar mass is 285.4 g/mol, find its molecular formula.

**Step 1.** Convert percentage composition to amount in mass (g).

| Elements: | C | : | H | : | N | : | O |
|---|---|---|---|---|---|---|---|
| Mass: | 71.54 g | | 6.72 g | | 4.91 g | | 16.82 g |

**Step 2.** Find the number of moles of each element present in the compound.

$$\frac{71.54\ \text{g}}{12.01\frac{g}{mol}\ \text{C}} \qquad \frac{6.72\ \text{g}}{1.01\frac{g}{mol}\ \text{H}} \qquad \frac{4.91\ \text{g}}{14.01\frac{g}{mol}\ \text{N}} \qquad \frac{16.82\ \text{g}}{16.00\frac{g}{mol}\ \text{O}}$$

**5.96 mol** : **6.66 mol** : **0.350 mol** : **1.05 mol**

**Step 3.** To get rid of decimals, divide all the numbers by the smallest one

$$\frac{5.96\ \text{mol}}{0.350} \qquad \frac{6.66\ \text{mol}}{0.350} \qquad \frac{0.35\ \text{mol}}{0.350} \qquad \frac{1.05\ \text{mol}}{0.350}$$

**17.0 mol** : **19.0 mol** : **1.0 mol** : **3.0 mol**

**Step 5.** To get the molecular formula for morphine, make these the subscripts of the specific elements in the molecule. Doing this we get $\mathbf{C_{17}H_{19}NO_3}$ for morphine.

**Molar mass check** $=17 \times 12.01$ g/mol C $+ 19 \times 10.1$ g/mol H $+ 1 \times 14.01$ g/mol N $+ 3 \times 16.0$ g /mol O $= 285.4$ g/mol

## *Problem:*

10. Viagara is one of a series of drugs that is used to treat men suffering from erectyle dysfunction. Its percentage composition by mass is: 55.67% C, 6.38% H, 17.71% N, 13.50 % O and 6.76% S. If its molar mass is 474.6 g/mol, find its molecular formula.

**Step 1.** Convert percentage composition to amount in mass (g).

| Elements: | C | : | H | : | N | : | O | : | S |
|---|---|---|---|---|---|---|---|---|---|
| Mass: | 55.67 g | | 6.38 g | | 17.71 g | | 13.50 g | | 6.76 g |

**Step 2.** Find the number of moles of each element present in the compound.

$$\frac{55.67\ \text{g}}{12.01\frac{g}{mol}\ \text{C}} \qquad \frac{6.38\ \text{g}}{1.01\frac{g}{mol}\ \text{H}} \qquad \frac{17.71\ \text{g}}{14.01\frac{g}{mol}\ \text{N}} \qquad \frac{13.50\ \text{g}}{16.00\frac{g}{mol}\ \text{O}} \qquad \frac{6.76\ \text{g}}{32.07\frac{g}{mol}\ \text{S}}$$

**4.63 mol C** : **6.32 mol H** : **1.27 mol N** : **0.844 mol O** : **0.211 mol S**

**Step 3.** To get rid of decimals, divide all the numbers by the smallest one, **0.211.**

$$\frac{4.64\ \text{mol}}{0.211} \qquad \frac{6.32\ \text{mol}}{0.211} \qquad \frac{1.26\ \text{mol}}{0.211} \qquad \frac{0.843\ \text{mol}}{0.211} \qquad \frac{0.211\ \text{mol}}{0.211}$$

**22 mol** : **30.0 mol** : **5.97 mol** : **3.99 mol** : **1.00 mol**

Rounding up the decimals and making these number the subscripts of the specific elements in the molecule. Doing this we get $\mathbf{C_{22}H_{30}N_6O_4S}$ for viagra.

Molar mass check $= 22 \times 12.01$ g $+ 30 \times 1.01$ g $+ 6 \times 14.01$ g $+ 4 \times 16$ g $+ 1 \times 32.07$ g $= 474.7$ g/mol

## *Problem:*

11. Atorvastatin (lipitor) is a member of a class of drugs used for lowering blood cholestrol. Its percentage composition by mass is: 70.95% C, 6.33% H, 3.40% F, 5.01% N, 14.29 % O. If its molar mass is 558.64 g/mol, find its molecular formula.

**Step 1.** Convert percentage composition to amount in mass (g).

| Elements: | C | : | H | : | F | : | N | : | O |
|---|---|---|---|---|---|---|---|---|---|
| Mass: | 70.95 g | | 6.33 g | | 3.40 g | | 5.01 g | | 14.29 |

**Step 2.** Find the number of moles of each element present in the compound.

$$\frac{70.95\ \text{g}}{12.01\frac{g}{mol}\ \text{C}} \qquad \frac{6.33\ \text{g}}{1.01\frac{g}{mol}\ \text{H}} \qquad \frac{3.40\ \text{g}}{\frac{19.00g}{mol}\ \text{F}} \qquad \frac{5.01\ \text{g}}{14.01\frac{g}{mol}\ \text{N}} \qquad \frac{14.29\ \text{g}}{16.00\frac{g}{mol}\ \text{O}}$$

**5.91 mol** : **6.26 mol** : **0.179 mol** : **0.358 mol** : **0.893 mol**

**Step 3.** To get rid of decimals, divide all the numbers by the smallest one, **0.179.**

$$\frac{5.91 \text{ mol}}{0.179} \qquad \frac{6.26 \text{ mol}}{0.179} \qquad \frac{0.179 \text{ mol}}{0.179} \qquad \frac{0.358 \text{ mol}}{0.179} \qquad \frac{0.893 \text{ mol}}{0.179}$$

**33.1 mol C  :    34.9 mol H  : 1.00 mol F   : 2.00 mol N : 4.99 mol O**

Rounding up the decimals and making these number the subscripts for the subscripts of the specific elements in the molecule. Doing this we get $C_{33}H_{35}F_1N_2O_5$ for Atorvastatin.

Molar mass check $= 33 \times 12.01 \text{ g} + 35 \times 1.01 \text{ g} \times 1 \times 19.00 \text{ g} +  \times 2 \times 14.01 \text{ g} + 5 \times 16.00 \text{ g}$

$= 558.64 \text{ g/mol}$

## *Problem:*

12. Caffeine, present in tea and coffee stimulates the central nervous system alleviating the symptoms fatigue and drowsiness; in premature babies it is used as a treatment to increase their lower than normal heart rates. Its percentage composition by mass is: 49.47% C, 5.20% H, 28.86% N,16.47 % O . If its molar mass is 194.19 g/mol, find its molecular formula.

Step 1. Convert percentage composition to amount in mass (g).

Elements:       C       :      H      :    N    :    O

Mass:       49.47 g      5.20 g      28.86 g      16.47 g

**Step 2.** Find the number of moles of each element present in the compound.

$$\frac{49.47 \text{ g}}{12.01 \frac{g}{mol} \, C} \qquad \frac{5.20 \text{ g}}{1.01 \frac{g}{mol} \, H} \qquad \frac{28.86 \text{ g}}{14.01 \frac{g}{mol} \, N} \qquad \frac{16.47 \text{ g}}{16.00 \frac{g}{mol} \, O}$$

**4.119 mol     :   5.15 mo   :   2.059 mol   :   1.029 mol**

**Step 3.** To get rid of decimals, divide all the numbers by the smallest one.

$$\frac{4.119 \text{ mol}}{1.029} \qquad \frac{5.15 \text{ mol}}{1.029} \qquad \frac{2.059 \text{ mol}}{1.029} \qquad \frac{1.029 \text{ mol}}{1.029}$$

**4 mol C   :    5 mol H   :    2 mol N    :    1 mol O**

**Step 4.** To get the empirical formula for caffeine, make these the subscripts of the specific elements in the molecule. Doing this we get $C_4H_5N_2O$ for the empirical formula of caffeine.

The molar mass for its empirical formula is:

$M = 4 \times 12.01 \text{ g/mol C} + 5 \times 1.01 \text{ g/mol H} + 2 \times 14.01 \text{ g/mol N} + 1 \times 16.00 \text{ g/mol}$

      48.04 g     +     5.05 g     +     28.02 g    + 16.00 g   *= 97.11 g/mol*

**Step 5.** Divide the molar mass of the compound by the molar mass of its empirical formula.

$$194.19 \text{ g/mol} \div 97.11 \text{ g/mol} = 2$$

**Step 6.** Multiply the subscript of each of the elements in the empirical formula by **2** to get the molecular formula for caffeine.

$$= C_{4 \times 2} H_{5 \times 2} N_{2 \times 2} O_{1 \times 2} = \; C_8H_{10}N_4O_2$$

## *Problem:*

13. Combustion of a sample of a compound containing only carbon and hydrogen produced 4.81g of carbon dioxide and 1.94 g of water. If the molar mass of the compound is 70.0 g/mol, calculate its molecular formula.

**Step 1.** Write an equation for the combustion of the hydrocarbon, using the letters x and as subscripts for carbon and hydrogen in the compound.

$$C_xH_y \ + \ O_{2(g)} \ \rightarrow \ CO_2 + H_2O \ + energy$$

**Step 2.** Find the amount of carbon in 4.81 g of carbon dioxide; this is what came from the sample of the compound that was burnt.

To do this, first find the percentage composition of carbon in carbon dioxide.

Molar mass of carbon dioxide = $1 \times 12.01$ g/mol C + $2 \times 16.00$ g/mol O = ***44.01g/mol***

$$\% \ C = \frac{12.01g}{44.01g} \times 100 \ \% \ = \mathbf{27.28 \ \%}$$

Mass of carbon in 4.81 g of $CO_2$ = 4.81 g/100 % $\times$ 27.28 % = **1.31 g**

$$C_xH_y \ + \ O_{2(g)} \ \rightarrow \ CO_2 + H_2O \ + energy$$

**Step 3.** Find the amount of hydrogen in **1.94 g** of water; this is what came from the sample of the compound that was burnt.

To do this, first find the percentage composition of hydrogen in water.

Molar mass of water = $2 \times 1.01$ g/mol H + $1 \times 16.00$ g/mol O = ***18.02g/mol***

$$\% \ H = \frac{2.02g}{18.02g} \times 100 \ \% \ = \mathbf{11.21 \ \%}$$

Mass of hydrogen in 1.94 g of $H_2O$ = 1.94 g /100 % $\times$ 11.2 % = **0.217 g**

**Step 4.** Find the mole ratio of C and H in the compound in the sample; 1.31 g of C and 0.217 g of H.

$$\mathbf{C} \qquad\qquad\qquad\qquad\qquad \mathbf{H}$$

$$\frac{1.31 \ g \ C}{12.01 \ g\frac{}{mol}C} = \ \mathbf{0.109 \ mol} \qquad : \qquad \frac{0.217 \ g \ H}{1.01 \ g\frac{}{mol \ H}} = \mathbf{0.215 \ mol}$$

**Step 5.** To get rid of decimals, divide all the numbers by the smallest one, **0.109.**

$$\frac{0.109}{0.109} = \mathbf{1} \qquad\qquad : \qquad\qquad \frac{0.217}{0.109} = \mathbf{2}$$

**Step 6.** To get the molecular formula for the compound, make these the subscripts of the specific elements in the molecule. Doing this, we get $C_1H_2$.

**Step 7.** Find the molar mass for its empirical formula.

$$1 \times 12.01 \ g/mol \ C \ + 2 \times 1.01 \ g/mol \ H = \textbf{\textit{14.02 g/mol}}$$

**Step 8.** Divide the molar mass of the compound by the molar mass of its empirical formula.

$$70.0 \ g/mol \div 14.02 \ g/mol \ = \mathbf{5}$$

**Step 9.** To get the molecular formula for the compound, multiply the subscripts of the elements in the empirical formula by **5.** Doing this we get

$$C_{1 \times 5}H_{2 \times 5} \ = \mathbf{C_5H_{10}}$$

# Stoichiometry



(a) $2HCl_{(aq)} + Na_2CO_{3(aq)} \rightarrow 2NaCl_{(aq)} + H_2O_{(l)} + CO_{2(g)}$

**4 mol**       **2 mol**       **4 mol**       **2 mol**       **2 mol**

(b) $2HCl_{(aq)} + Na_2CO_{3(aq)} \rightarrow 2NaCl_{(aq)} + H_2O_{(l)} + CO_{2(g)}$

**1 mol**       **½ mol**       **1 mol**       **½ mol**       **½ mol**

(c) $H_2SO_{4(aq)} + 2\,NaOH_{(aq)} \rightarrow 2\,NaCl_{(aq)} + 2\,H_2O_{(l)}$

**½ mol**       **1 mol**       **1 mol**       **1 mol**

(d) $H_2SO_{4(aq)} + 2\,NaOH_{(aq)} \rightarrow 2\,NaCl_{(aq)} + 2\,H_2O_{(l)}$

**¼mol**       **½ mol**       **½ mol**       **½ mol**

(e) $Ba(OH)_{2(aq)} + 2\,HNO_{3(aq)} \rightarrow Ba(NO_3)_{2(aq)} + 2\,H_2O_{(l)}$

**2 mol**       **4 mol**       **2 mol**       **4 mol**

(f) $Ba(OH)_{2(aq)} + 2HNO_{3(aq)} \rightarrow Ba(NO_3)_{2(aq)} + 2H_2O_{(l)}$

**1/3 mol**       **2/3 mol**       **1/3 mol**       **2/3 mol**

(g) $2\,BrO_{3\,(aq)}^- + 5\,HSO_{3\,(aq)}^- \rightarrow Br_{2(aq)} + 5\,SO_{4\,(aq)}^{2-} + H_2O_{(l)} + 3\,H_{(aq)}^+$

**3 mol**       **7.5 mol**       **1.5 mol**       **7.5 mol**       **1.5 mol**       **4.5 mol**

(h) $2\,BrO_{3\,(aq)}^- + 5\,HSO_{3\,(aq)}^- \rightarrow Br_{2(aq)} + 5\,SO_{4\,(aq)}^{2-} + H_2O_{(l)} + 3\,H_{(aq)}^+$

**0.8 mol**       **2 mol**       **0.4 mol**       **2 mol**       **0.4 mol**       **1.2 mol**

(i) $3\,NO_{2(g)} + H_2O_{(l)} \rightarrow 2\,HNO_{3(aq)} + NO_{(g)}$

**9 mol**       **3 mol**       **6 mol**       **3 mol**

(j) $3\,NO_{2(g)} + H_2O_{(l)} \rightarrow 2\,HNO_{3(aq)} + NO_{(g)}$

**4 ½ mol**       **1 ½ mol**       **3 mol**       **1 ½ mol**



*Problem:*

       Given         Required

**(a)**   $2\,AgNO_{3(aq)} + BaCl_{2(aq)} \rightarrow 2\,AgCl_{(s)} + Ba(NO_3)_{2(aq)}$

      84.94 g            ?

*What is Required?*

The mass of **$BaCl_2$** that is required to react with **84.94 g** of **$AgNO_3$**.

*Provided?*

   Mass of $AgNO_3 = 84.94$ g

*Steps to be taken:*

First, find the molar mass of **$AgNO_3$** and then convert **84.94 g** of it to number of moles. Then use the number of moles of **$AgNO_3$** to find the number of **$BaCl_2$** that is required to react with it, using the mole ratios in the equation. Finally, convert the number of moles of **$BaCl_2$** to amount in mass.

**Step 1.** Molar mass for **$AgNO_3$** = $1 \times 107.87$ g/mol Ag $+\ 1 \times 14.01$ g/mol N $+\ 3 \times 16.00$ g/mol O

          = **169.9 g/mol**

**Step 2.** Convert **84.94 g** of **$AgNO_3$** to number of moles.

    Using the formula, $n = \dfrac{m}{M} = \dfrac{84.94\ \text{g}}{169.9\ \text{g/mol}} = 0.5000$ mol

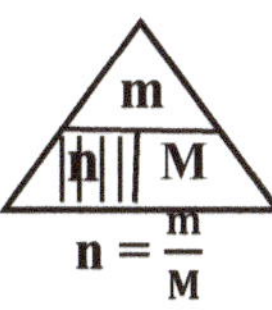

**Step 3.** Find the number of moles of **BaCl₂** that are required to react with **0.5000 mol** of **AgNO₃**, using the mole ratios in the balanced equation.

$$\frac{\text{Given}}{\text{Required}} = \frac{\textbf{2 mol}}{\textbf{1 mol}} \diagdown \frac{0.5000 \text{ mol AgNO3}}{\text{X mol BaCl2}} = 2X = \textbf{0.5000 mol BaCl}_2$$

$$X = \frac{0.5000 \text{ mol}}{2} = \textbf{0.2500 mol BaCl}_2$$

**Step 4.** Convert the number of moles of **BaCl₂** to amount in mass. To do this, first find the molar mass of **BaCl₂**.
Molar mass = $1 \times 137.33$ g/mol Ba $+ 2 \times 35.45$ g/mol Cl = ***208.23 g/mol***

Using the formula, **m = n × M we get**

Mass = $0.2500$ mol $\times 208.23$ g/mol BaCl₂

= **52.06 g BaCl₂**

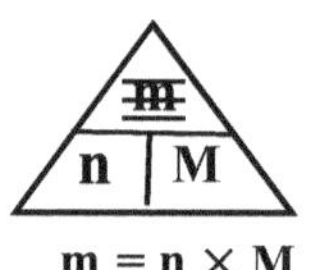


## *Problem:*

| Required | Given |
| --- | --- |
| | |

**(b)** $CuSO_{4(aq)}$ $+$ $2\,NaOH_{(aq)}$ $\rightarrow$ $Cu(OH)_{2(s)}$ $+$ $Na_2SO_{4(aq)}$
 $\quad$ ? $\qquad\qquad$ 8.00 g

*What is Required?*

The mass of **CuSO₄** that is required to react with **8.00 g** of **NaOH**.

*Provided?*

Mass of NaOH = 8.00 g

*Steps to be taken:*

First, find the molar mass of **NaOH** and then convert **8.00 g** of it to number of moles. Then use the number of moles of **NaOH** to find the number of moles of **CuSO₄** that are required to react with it, using the mole ratio in the equation. Finally, convert the number of moles of **CuSO₄** to amount in mass.

**Step 1.** Molar mass for **NaOH** = $1 \times 22.99$ g/mol Na$+ 1 \times 16.00$ g/mol O $+ 1 \times 12.01$ g/mol H

= **40.00 g/mol**

**Step 2.** Convert **8.0 g** of **NaOH** to number of moles.

Using the formula, $\mathbf{n = \dfrac{m}{M}}$ we get $\mathbf{n} = \dfrac{8.00 \text{ g}}{40.00 \text{ g/mol}} = \textbf{0.200 mol}$

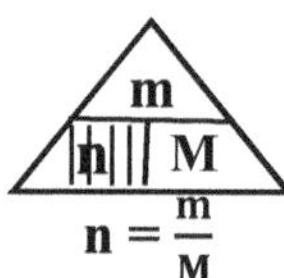


**Step 3.** Find the number of moles of **CuSO₄** that is required to react with **0.200 mol** of **NaOH**, using the mole ratio in the balanced equation.

$$\frac{\text{Given}}{\text{Required}} = \frac{\textbf{2 mol}}{\textbf{1 mol}} \diagdown \frac{0.200 \text{ mol NaOH}}{\text{X mol CuSO}_4} = 2X = \textbf{0.200 mol CuSO}_4$$

$$X = \frac{0.200 \text{ mol}}{2} = \textbf{0.100 mol CuSO}_4$$

**Step 4.** Convert the number of moles of **CuSO₄** to amount in mass. To do this, first find its molar mass.

Molar mass = $1 \times 63.55$ g/mol Cu $1 \times 32.07$ g/mol S $+ 4 \times 16.00$ g/mol O = ***159.6 g/mol***

Using the formula, **m = n × M we get**

Mass = $0.100$ mol $\times 159.6$ g/mol = **15.9 g CuSO₄**

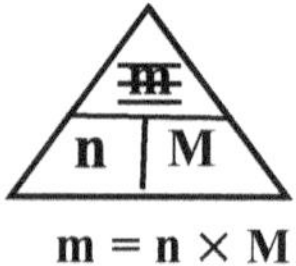

*Problem:*

Required      Given

**(c)**   $Mg_{(s)}$    +   $CuSO_{4(aq)}$   $\rightarrow$   $MgSO_{4(aq)}$   +   $Cu_{(s)}$

      ?          15.96 g

*What is Required?*

The mass of $Mg_{(s)}$ that is required to react with **15.96 g** of **CuSO₄**

*Provided?*

**15.96 g** of **CuSO₄**

*Steps to be taken:*

First, find the molar mass of **CuSO₄** and then convert **15.96 g** of it to number of moles. Then use the number of moles of **CuSO₄** to find the number of moles of **Mg** that is required to react with it, using the mole ratios in the equation. Finally convert the number of moles of **Mg** to amount in mass.

**Step 1.** Molar mass for $CuSO_4 = 1 \times 63.55$ g/mol Cu $+ 1 \times 32.07$ g/mol S $+ 4 \times 16.00$ g/mol O

$$= \textbf{159.6 g/mol}$$

**Step 2.** Convert **15.96 g** of **CuSO₄** to number of moles

Using the formula, $\mathbf{n = \dfrac{m}{M}}$ we get $\mathbf{n = \dfrac{15.96 \text{ g}}{159.6 \text{ g/mol}} = 0.1000 \text{ mol}}$

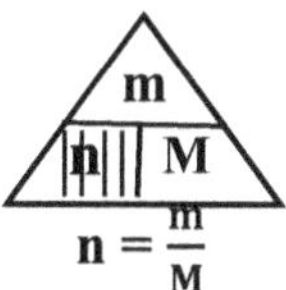

**Step 3.** Find the number of moles of **Mg** that is required to react with **0.1000 mol** of **CuSO₄**, using the mole ratios in the balanced equation.

$$\frac{\text{Given}}{\text{Required}} = \frac{1 \text{ mol}}{1 \text{ mol}} = \frac{0.1000 \text{ mol CuSO}_4}{\text{X mol Mg}} = X = \textbf{0.1000 mol Mg}$$

**Step 4.** Convert the number of moles of **Mg** to amount in mass. To do this, first find the molar mass of **Mg**. Molar mass $= 1 \times 24.31$ g/mol Mg $= \textbf{24.31 g/mol}$

Using the formula, $\mathbf{m = n \times M}$ we get

Mass $= 0.1000$ mol $\times 24.31$ g/mol Mg $= \textbf{2.431g}$

$$m = n \times M$$

*Problem:*

Required      Given

**(d)**   $Al(OH)_{3(s)}$   +   $3 \, HCl_{(aq)}$    $\rightarrow$   $3 \, AlCl_{3(aq)}$   +   $3 \, H_2O_{(l)}$

      ?          18.23 g

*Required:*

The mass of $Al(OH)_3$ that is required to react with **18.23 g** of **HCl.**

*Provided:* Mass of HCl $= 18.23$ g

*Steps to be taken:*

First, find the molar mass of **HCl** and then convert **18.23 g** of it to number of moles. Then use the number of moles of **HCl** to find the number of moles of $Al(OH)_3$ that is required to react with it, using the mole ratios in the equation. Finally, convert the number of moles of $Al(OH)_3$ to amount in mass.

**Step 1.** Molar mass for **HCl** = 1 × 1.01 g/mol H + 1 × 35.45 g/mol Cl = *36.45 g/mol HCl*

**Step 2.** Convert **18.23 g** of **HCl** to number of moles

Using the formula, $n = \dfrac{m}{M}$ we get $n = \dfrac{18.23\ g}{36.45\ g/mol} = $ **0.5000 mol HCl**

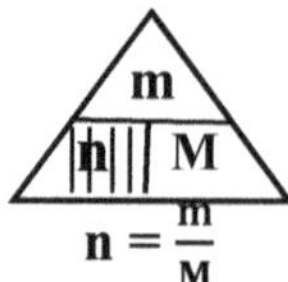

**Step 3.** Find the number of moles of **Al(OH)₃** that are required to react with **0.5000 mol** of **HCl**, using the mole ratio in the balanced equation.

$$\frac{Given}{Required} = \frac{3\ mol}{1\ mol} \quad\times\quad \frac{0.5000\ mol\ HCl}{X\ mol\ Al(OH)_3} = 3X = 0.5000\ mol\ Al(OH)_3$$

$$X = \frac{0.5000\ mol}{3} = 0.1667\ mol\ Al(OH)_3$$

**Step 4.** Convert the number of moles of **Al(OH)₃** to amount in mass. To do this, first find its molar mass.

Molar mass = 1 × 26.98 g/mol Al + 3 × 1.01 g/mol H + 3 × 16.00 g/mol O = *78.01 g/mol*

**Step 5.** Find the mass of **0.1667 mol Al(OH)₃.**

Using the formula, $m = n \times M$ we get

Mass = 0.1667 mol × 78.01 g/mol = **13.00 g Al(OH)₃**

## *Problem:*

|  | Given |  | Required |  |  |  |  |
|---|---|---|---|---|---|---|---|
| (e) | $Zn_{(s)}$ | + | $H_2SO_{4(aq)}$ | → | $ZnSO_4$ | + | $H_{2(g)}$ |
|  | 16.35 g |  | ? |  |  |  |  |

*Required:*

The mass of **H₂SO₄** that is required to react with **16.35 g** of **Zn.**

*Provided:*

Mass of Zn = 16.35 g

*Steps to be taken:*

First convert **16.35 g** of **Zn** to amount in moles. Then use the number of moles of **Zn** found, to find the number of moles of **H₂SO₄₍ₐq₎** that are required to react with it, using the mole ratio in the balanced equation. Finally, convert the number of moles of **H₂SO₄₍ₐq₎** to amount in mass.

**Step 1.** Molar mass for **Zn** = 1 × 65.39 g/mol = **65.39 g/mol**

**Step 2.** Convert **16.35 g** of **Zn** to number of moles

Using the formula, $n = \dfrac{m}{M} = \dfrac{16.35\ g}{65.39\ g/mol} = $ **0.2500 mol Zn**

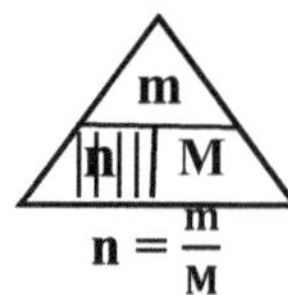

**Step 3.** Find the number of moles of **H₂SO₄** that are required to react with **0.2500 mol** of **Zn** using the mole ratio in the balanced equation.

$$\frac{Given}{Required} = \frac{1\ mol}{1\ mol} \quad\times\quad \frac{0.2500\ mol\ Zn}{X\ mol\ H_2SO_4} = X = 0.2500\ mol\ H_2SO_4$$

**Step 4.** Convert the number of moles of $H_2SO_4$ to amount in mass. To do this, first find its molar mass. M = $2 \times 1.01$ g/mol H + $1 \times 32.07$ g/mol S + $4 \times 16.00$ g/mol O = ***98.09 g/mol***

Using the formula, **m = n × M we get**

Mass = 0.2500 mol × 98.09 g/mol = **24.52 g** $H_2SO_{4(aq)}$

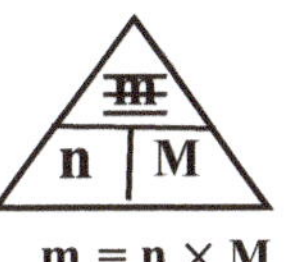

## Problems involving excess and limiting reagents

**Exercise 11.3 (p.180)**

***Problem:***  Given

(a)  $H_2SO_{4(aq)}$  +  2 NaOH$_{(aq)}$  →  2 NaCl$_{(aq)}$  +  2 $H_2O_{(l)}$

| 20.00 g | excess |  | ? | ? |

Required  Required

*Required:*

The masses of **NaCl** and **H₂O** produced

*Provided:*

Mass of $H_2SO_4$ = 20.00 g

An excess of NaOH

*Steps to be taken:*

Since one reagent, **(NaOH)** is **in excess,** the other one, **$H_2SO_4$** is the *limiting reagent* (**L.R**). The first step will be to find the number of moles of **$H_2SO_4$**, and then use this value to find the number of moles of **NaCl** and **H₂O** produced, using the mole-ratios in the balanced equation. Finally, convert their amounts in moles to amounts in mass.

**Step 1.** Find the number of moles of **$H_2SO_4$** in **20.00 g.** To do this, first find its molar mass.

M = $2 \times 1.01$ g/mol H + $1 \times 32.07$ g/mol S + $4 \times 16.00$ g/mol O = ***98.09 g/mol***

Using the formula, $n = \dfrac{m}{M}$ we get $n = \dfrac{20.00 \text{ g}}{98.09 \text{ g/mol}}$

= **0.2038 mol (L.R)**

**Step 2.** Find the number of moles of **NaCl**, using the number of moles of the limiting reagent.

$$\frac{\text{Given}}{\text{Required}} = \frac{1 \text{ mol}}{2 \text{ mol}} = \frac{0.2038 \text{ mol } H_2SO_4}{X \text{ mol NaCl}} \quad X = \textbf{0.4076 mol NaCl}$$

**Step 3.** Find the mass of 0.4076 mol **NaCl.** To do this, first find its molar mass.

M = $1 \times 22.99$ g/mol Na + $1 \times 35.45$ g/mol Cl = ***58.44 g/mol***

Using the formula, **m = n × M** we get

Mass = 0.4076 mol × 58.44 g/mol = **23.82 g NaCl**

**Step 4.** Find the number of moles of **H₂O**, using the number of moles of the limiting reagent.

$$\frac{\text{Given}}{\text{Required}} = \frac{1 \text{ mol}}{2 \text{ mol}} = \frac{0.2038 \text{ mol } H_2SO_4}{X \text{ mol } H_2O} \quad X = \textbf{0.4076 mol } H_2O$$

**Step 5.** Find the mass of **0.4076 mol H$_2$O.** To do this, first find its molar mass.

$$M = 2 \times 1.01 \text{ g/mol H} + 1 \times 16.00 \text{ g/mol O} = \textit{18.02 g/mol}$$

Using the formula, $\mathbf{m = n \times M}$ we get

$$\text{Mass} = 0.4076 \text{ mol} \times 18.02 \text{ g/mol} = \textbf{7.34 g H}_2\textbf{O}$$

## *Problem:*

|  | | Given | | Required | | Required |
|---|---|---|---|---|---|---|
| (b)  Ba(OH)$_2$ | + | 2 HNO$_{3(aq)}$ | $\rightarrow$ | Ba(NO$_3$)$_{2(aq)}$ | + | 2 H$_2$O$_{(l)}$ |
| excess | | 15.75 g | | ? | | ? |

*What is Required?*

The masses of **Ba(NO$_3$)$_2$** and **H$_2$O** produced.

*Provided:*

Mass of HNO$_{3(aq)}$ = 15.75 g

An excess of Ba(OH)$_2$

*Strategy:*

Since one reagent, **(Ba(OH)$_2$)** is in excess the other, **(HNO$_3$)** is limiting. The first step will be to find the number of moles of **HNO$_3$,** and then use this value to find the number of moles of **Ba(NO$_3$)$_2$** and **H$_2$O** produced, using the mole-ratios in the balanced equation. Finally, convert their amounts in moles to amounts in mass.

*Steps to be taken:*

**Step 1.** Find the number of moles of **HNO$_3$** in 15.75 g. To do this, first find its molar mass.

$$M = 1 \times 1.01 \text{ g/mol H} + 1 \times 14.01 \text{ g/mol N} + 3 \times 16.00 \text{ g/mol O} = \textit{63.02 g/mol}$$

Using the formula, $\mathbf{n = \dfrac{m}{M}}$ we get $\mathbf{n} = \dfrac{15.75 \text{ g}}{63.02 \text{ g/mol}}$

$$= \textbf{0.2499 mol HNO}_3 \textbf{ (L.R)}$$

**Step 2.** Find the number of moles of Ba(NO$_3$)$_2$, using the number of moles of the limiting reagent.

$$\dfrac{\text{Given}}{\text{Required}} = \dfrac{2 \text{ mol}}{1 \text{ mol}} \quad \dfrac{0.2499 \text{ mol } \textbf{HNO}_3}{\text{X mol } \textbf{Ba(NO}_3\textbf{)}_2}$$

$$\textbf{2 X = 0.2499 mol} = \text{X} = \textbf{0.1249 mol Ba(NO}_3\textbf{)}_2$$

**Step 3.** Find the mass of **0.1249 mol Ba(NO$_3$)$_2$** . To do this, first find its molar mass.

$$M = 1 \times 137.33 \text{ g/mol Ba} + 2 \times 14.01 \text{ g/mol N} + 6 \times 16.00 \text{ g/mol O} = \textit{261.35 g/mol}$$

Using the formula, $\mathbf{m = n \times M}$ we get

$$\text{Mass} = 0.1249 \text{ mol} \times 261.35 \text{ g/mol} = \textbf{32.64 g Ba(NO}_3\textbf{)}_2$$

**Step 4.** Find the number of moles of **H$_2$O**, using the number of moles of the limiting reagent.

$$\dfrac{\text{Given}}{\text{Required}} = \dfrac{2 \text{ mol}}{2 \text{ mol}} \quad \dfrac{0.1249 \text{ mol} \textbf{HNO}_3}{\text{X mol } \textbf{H}_2\textbf{O}}$$

$$\textbf{2 X = 0.2499 mol} = \text{X} = \textbf{0.1249 mol H}_2\textbf{O}$$

**Step 5**. Find the mass of **0.1249 mol H$_2$O**. To do this, first find its molar mass.

$$M = 2 \times 1.01 \text{ g/mol H} + 1 \times 16.00 \text{ g/mol O} = \textbf{\textit{18.02 g/mol}}$$

Using the formula, **m = n $\times$ M** we get

$$\text{Mass} = 0.1249 \text{ mol} \times 18.02 \text{ g/mol} = \textbf{2.251 g H}_2\textbf{O}$$

## *Problem:*

| | Given | | | Required | | Required |
|---|---|---|---|---|---|---|
| (c) | 3 NO$_{2(g)}$ | + | H$_2$O$_{(l)}$ $\rightarrow$ | 2 HNO$_{3(aq)}$ | + | NO$_{(g)}$ |
| | 23.00 g | | excess | ? | | ? |

*Required:*

The masses of HNO$_3$ and NO

*Provided:*

Mass of NO$_{2(g)}$ = 23.00 g

An excess of H$_2$O$_{(l)}$

*Steps to be taken*:

Since one reagent, **H$_2$O** is in excess, the other one, **NO$_2$** is limiting. The first step will be to find the number of moles of **NO$_2$** and then use this value to find the number of moles of **HNO$_3$** and **NO** produced, using the mole-ratios in the balanced equation. Finally, convert their amounts in moles to amounts in mass.

**Step 1.** Find the number of moles of **NO$_2$** in **23.00 g**. To do this, first find its molar mass.

$$M = 1 \times 14.01 \text{ g/mol N} + 2 \times 16.00 \text{ g/mol O} = \textbf{\textit{46.01 g/mol}}$$

Using the formula, $\mathbf{n = \dfrac{m}{M}}$ we get $\mathbf{n = \dfrac{23.00 \text{ g}}{46.01 \text{ g/mol}}}$

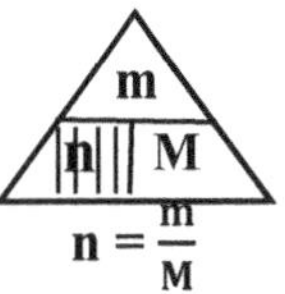

$$= \textbf{0.5000 mol NO}_2 \text{ (L.R)}$$

**Step 2.** Find the number of moles of **HNO$_3$**, using the number of moles of the limiting reagent.

$$\frac{\text{Given}}{\text{Required}} = \frac{3 \text{ mol}}{2 \text{ mol}} \diagdown\!\!\!\!\diagup \frac{0.5000 \text{ mol NO}_2}{\text{X mol HNO}_3}$$

$$\textbf{3 X = 1.000 mol HNO}_3 = \textbf{X = 0.333 mol HNO}_3$$

**Step 3.** Find the mass of **0.333 mol HNO$_3$**. To do this, first find its molar mass.

$$M = 1 \times 1.01 \text{ g/mol H} + 1 \times 14.01 \text{ g/mol N} + 3 \times 16.00 \text{ g/mol O} = \textbf{\textit{63.0 g/mol}}$$

Using the formula, **m = n $\times$ M** we get

$$\text{Mass} = 0.333 \text{ mol} \times 63.00 \text{ g/mol} = \textbf{21.0 g HNO}_3$$

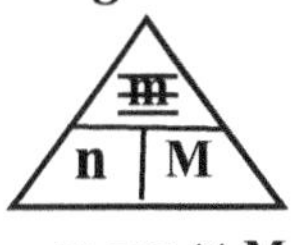

**Step 4.** Find the number of moles of **NO**, using the number of moles of the limiting reagent.

$$\frac{\text{Given}}{\text{Required}} = \frac{3 \text{ mol}}{1 \text{ mol}} \diagdown\!\!\!\!\diagup \frac{0.333 \text{ mol NO}_2}{\text{X mol NO}}$$

$$\textbf{3 X = 0.333 mol NO} = \textbf{X = 0.111 mol NO}$$

**Step 5.** Find the mass of **0.111 mol NO**. To do this, first find its molar mass.

$$M = 1 \times 14.01 \text{ g/mol N} + 1 \times 16.00 \text{ g/mol O} = \textbf{\textit{30.01 g/mol}}$$

Using the formula, **m = n $\times$ M** we get

$$\text{Mass} = 0.111 \text{ mol} \times 30.01 \text{ g/mol} = \textbf{3.33 g NO}$$

*Problem:*

| Given | | Required | Required | Required |
|---|---|---|---|---|

(d) $CaCO_{3(s)}$ + 2 $HCl_{(aq)}$ → $CaCl_{2(aq)}$ + $H_2O_{(l)}$ + $CO_{2(g)}$
   excess            9.12 g            ?            ?            ?

*Provided:*

Mass of $HCl_{(aq)}$ = 9.12 g

An excess of $CaCO_{3(s)}$

*Required:*

The masses of **$CaCl_2$, $H_2O$ and $CO_2$** produced

*Steps to be taken:*

Since one reagent, **$CaCO_3$** is *in excess,* the other one, **HCl** *is limiting.* The first step will be to find the number of moles of **HCl** and then use this value to find the number of moles of **$CaCl_2$, $H_2O$ and $CO_2$** produced, using the mole-ratios in the balanced equation. Finally, convert their amounts in moles to amounts in mass.

**Step 1**. Find the number of moles of **HCl** in **9.12 g.** To do this, first find its molar mass.

$$M = 1\times 1.01 \text{ g/mol H} + 1\times 35.45 \text{ g/mol Cl} = \textbf{\textit{36.46 g/mol}}$$

Using the formula, $n = \dfrac{m}{M}$ we get = $n = \dfrac{9.12 \text{ g}}{36.46 \text{ g/mol}}$

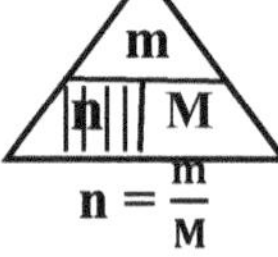


$$= \textbf{0.250 mol HCl (L.R)}$$

**Step 2.** Find the number of moles of **$CaCl_2$**, using the number of moles of the limiting reagent.

$$\frac{\text{Given}}{\text{Required}} = \frac{2 \text{ mol}}{1 \text{ mol}} \quad \frac{0.250 \text{ mol } \textbf{HCl}}{\text{X mol } \textbf{CaCl}_2}$$

$$\textbf{2 X} = \textbf{0.250 mol CaCl}_2 = \textbf{X} = \textbf{0.125 mol CaCl}_2$$

**Step 3.** Find the mass of **0.125 mol $CaCl_2$.** To do this, first find its molar mass.

$$M = 1 \times 40.8 \text{ g/mol Ca} + 2 \times 35.45 \text{ g/mol Cl} = \textbf{\textit{111.0 g/mol}}$$

Using the formula, $m = n \times M$ we get

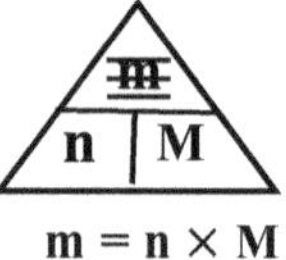


Mass = 0.125 mol × 111.0 g/mol = **13.9 g $CaCl_2$**

**Step 4.** Find the number of moles of **$H_2O$**, using the number of moles of the limiting reagent.

$$\frac{\text{Given}}{\text{Required}} = \frac{2 \text{ mol}}{1 \text{ mol}} \quad \frac{0.250 \text{ mol } \textbf{HCl}}{\text{X mol } \textbf{H}_2\textbf{O}}$$

$$\textbf{2 X} = \textbf{0.250 mol H}_2\textbf{O} = \textbf{X} = \textbf{0.125 mol H}_2\textbf{O}$$

**Step 5.** Find the mass of **0.125 mol $H_2O$.** To do this, first find its molar mass.

$$M = 2 \times 1.01 \text{ g/mol H} + 1 \times 16.00 \text{ g/mol O} = \textbf{\textit{18.02 g/mol}}$$

Using the formula, $m = n \times M$ we get

Mass = 0.125 mol × 18.02 g/mol = **2.25 g $H_2O$**

**Step 6.** Find the number of moles of $CO_2$, using the number of moles of the limiting reagent.

$$\frac{\text{Given}}{\text{Required}} = \frac{2 \text{ mol}}{1 \text{ mol}} \quad \diagup\!\!\!\!\diagdown \quad \frac{0.250 \text{ mol } \mathbf{HCl}}{X \text{ mol } \mathbf{CO_2}}$$

$$2X = 0.250 \text{ mol } CO_2 = X = \mathbf{0.125 \text{ mol } CO_2}$$

**Step 7.** Find the mass of **0.125 mol $CO_2$**. To do this, first find its molar mass.

$$M\ CO_2 = 1 \times 12.01 \text{ g/mol C} + 2 \times 16.00 \text{ g/mol O} = \mathbf{44.01 \text{ g/mol}}$$

Using the formula, $\mathbf{m = n \times M}$ we get

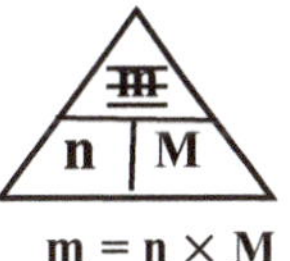


$$\text{Mass} = 0.125 \text{ mol} \times 44.01 \text{ g/mol} = \mathbf{5.50 \text{ g } CO_2}$$

## Exercise 11.4 (p.190)

***Problem:***

**1.** Find the mass of $H_{2(g)}$ produced by the reaction between 6.34 g $Zn_{(s)}$ and has 4.90 g $H_2SO_{4(aq)}$.

*Provided:*

Mass of Zn = 6.34 g

Mass of $H_2SO_4$ = 4.90 g

*Required:*

The mass of hydrogen produced

*Steps to be carried out:*

**Step 1.** Write a balanced equation for the reaction

$$Zn_{(s)} \quad + \quad H_2SO_{4(aq)} \quad \rightarrow \quad ZnSO_{4(aq)} + \quad H_{2(g)}$$

First is necessary to find which of the two reactants, Zn or $H_2SO_4$ is the limiting reagent, and then use it to find the mass of hydrogen produced. To do this, the following other steps are carried out.

**Step 2.** Find the number of moles **Zn**. To do this, first find its molar mass.

$$M = 1 \times 65.39 \text{ g/mol Zn} = \mathbf{65.39 \text{ g/mol}}$$

Using the formula, $\mathbf{n = \dfrac{m}{M}}$ we get $\mathbf{n} = \dfrac{6.34 \text{ g}}{65.39 \dfrac{g}{mol} Zn}$

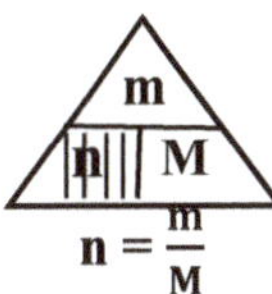


$$= \mathbf{0.0969 \text{ mol Zn}} \ \textit{(Used or Experimental value)}$$

**Step 3.** Find the number of moles **$H_2SO_4$**. To do this, first find its molar mass.

$$M = 2 \times 1.01 \text{ g/mol H} + 1 \times 32.07 \text{ g/mol S} + 4 \times 16.00 \text{ g/mol O} = \mathbf{98.09 \text{ g/mol}}$$

Using the formula, $\mathbf{n = \dfrac{m}{M}}$ we get $\mathbf{n} = \dfrac{4.90 \text{ g}}{98.09 \text{ g/mol}}$

$$= \mathbf{0.0500 \text{ mol } H_2SO_4} \ \textit{(Used or experimental value)}$$

**Step 4.** Find the limiting reagent. To do this, choose any one of the two reactants and determine how much of the other, in theory, is required to react with it. We choose **0.0500 mol H₂SO₄** to see how much **Zn** is required to react with it. In this case, the **H₂SO₄** becomes the *given reagent* and **Zn** becomes the *required reagent.*

$$\frac{Given}{Required} = \frac{1\ mol}{1\ mol} \diagdown \diagup \frac{0.0500\ mol\ \mathbf{H_2\ SO_4}}{X\ mol\ \mathbf{Zn}}$$

$$X = \mathbf{0.0500\ mol\ Zn}\ (\textit{Theoretical or required value})$$

$$\mathbf{0.0969\ mol\ Zn}\ (\textit{Experimental value})$$

Comparison of these two values shows that **more Zn (0.0969 mol)** was used than was required **(0.050 mol)**. It therefore becomes the *excess reagent* and **H₂SO₄** becomes the *limiting reagent.*

Excess amount of **Zn** = 0.0969 mol - 0.050 mol = **0.0469 mol**

**Step 4.** Use the limiting reagent, **H₂SO₄** to calculate the amount of **H₂(g)** produced.

$$\frac{Given}{Required} = \frac{1\ mol}{1\ mol} \diagdown \diagup \frac{0.0500\ mol\ \mathbf{H_2\ SO_4}}{X\ mol\ \mathbf{H_2}}$$

$$X = \mathbf{0.050\ mol\ H_{2(g)}}$$

**Step 5.** Find the mass of **0.0500 mol H₂(g)**

Mass = **0.0500 mol** × **2.02 g/mol H₂** = **0.101 g H₂**

## *Problem:*

**2.** Calculate the mass of $Cu(OH)_{2(s)}$ produced from a reaction between 15.96 g $CuSO_{4(aq)}$ and 4.00 g NaOH(aq).

*Provided:*

Mass of $CuSO_4$ = 15.96 g

Mass of NaOH = 4.00 g

*Required:*

The mass of $Cu(OH)_{2(s)}$ produced

*Steps to be carried out:*

**Step 1.** Write a balanced equation for the reaction.

$$CuSO_{4(aq)} \quad + \quad 2\ NaOH_{(aq)} \quad \longrightarrow \quad Cu(OH)_{2(s)} \ + \ 2\ H_2O_{(l)}$$

First, is necessary to find which of the two reactants, $CuSO_{4(aq)}$ or $NaOH_{(aq)}$ is the limiting reagent, then use it to find the mass of $Cu(OH)_{2(s)}$ produced. To do this, the following steps are carried out.

**Step 2.** Find the number of **moles CuSO₄ in 15.96 g**. To do this, first find its molar mass.

M = 1 × 63.55 g/mol Cu+ 1 × 32.07 g/mol S + 4 × 16.00 g/mol O = *159.6 g/mol*

Using the formula, $\mathbf{n = \dfrac{m}{M}}$ we get $\mathbf{n = \dfrac{15.96\ g}{159.6\ g/mol}}$

$$= \mathbf{0.1000\ mol\ CuSO_4}\ (Experimental\ value)$$

$$n = \frac{m}{M}$$

**Step 3.** Find the number of **moles NaOH in 4.00 g**. To do this, first find its molar mass.

$$M = 1 \times 22.99 \text{ g/mol Na} + 1 \times 16.00 \text{ g/mol O} + 1 \times 1.01 \text{ g/mol H} = \boldsymbol{40.00 \text{ g/mol}}$$

Using the formula, $\mathbf{n = \dfrac{m}{M}}$ we get $\mathbf{n} = \dfrac{4.00 \text{ g}}{40.00 \text{ g/mol}}$

$$= \textbf{0.100 mol NaOH} \text{ (Experimental value)}$$

**Step 3.** Find the limiting reagent. To do this, choose anyone of the two reactants and determine how much of the other is required, in theory, is required to react with it. We choose **0.100 mol NaOH** to find how much **CuSO$_4$** is required to react with it. In this case, the **0.100 mol NaOH** becomes the *given reagent* and **CuSO$_4$** becomes the *required reagent.*

$$\frac{\text{Given}}{\text{Required}} = \frac{2 \text{ mol}}{1 \text{ mol}} \quad \times \quad \frac{0.100 \text{ mol NaOH}}{X \text{ mol CuSO}_4}$$

$$\textbf{2 X = 0.100 mol CuSO}_4$$

$$\textbf{X = 0.050 mol CuSO}_4 \text{ (Theoretical value)}$$

$$\textbf{0.1000 mol CuSO}_4 \text{ (Experimental value)}$$

Comparison of these two values shows that **more CuSO$_4$** was used (**0.1000 mol**) than was required (**0.050 mol**). It therefore becomes the *excess reagent* and **NaOH** becomes the *limiting reagent.*

The excess amount of **CuSO$_4$** = 0.100 mol – 0.050 mol = **0.050 mol**

**Step 4.** Use the limiting reagent, **NaOH** to calculate the amount of **Cu(OH)$_{2(s)}$** formed.

$$\frac{\text{Given}}{\text{Required}} = \frac{2 \text{ mol}}{1 \text{ mol}} \quad \times \quad \frac{0.100 \text{ mol NaOH}}{X \text{ mol Cu(OH)}_2}$$

$$\textbf{2 X = 0.100 mol Cu(OH)}_2 \quad \textbf{= X = 0.0500 mol Cu(OH)}_2$$

**Step 5.** Find the mass of **0.0500mol Cu(OH)$_{2(s)}$** formed. To do this, first find its molar mass.

$$M = 1 \times 63.55 \text{ g/mol Cu} + 2 \times 16.00 \text{ g/mol O} + 2 \times 1.01 \text{ g/mol H} = \boldsymbol{97.57 \text{ g/mol}}$$

Using the formula, $\mathbf{m = n \times M}$ we get

$$\text{Mass} = 0.0500 \text{ mol} \times 97.57 \text{ g/mol} = \textbf{4.88 g Cu(OH)}_{2(s)}$$

$$m = n \times M$$

*Problem:*

3. Calculate the masses of $CO_{2(g)}$ and $H_2O_{(l)}$ produced when 4.00 g $CH_{4(g)}$ is burnt in 8.00 g $O_{2(g)}$.

*Provided:*

Mass of $CH_4$ = 4.00 g

Mass of $O_{2(g)}$ = 8.00 g

*Required:*

The masses of $CO_{2(g)}$ and $H_2O_{(l)}$ produced

*Steps to be carried out:*

**Step 1.** Write a balanced equation for the reaction

$$CH_{4(g)} \quad + \quad 2\,O_{2(g)} \quad \longrightarrow \quad CO_{2(g)} \quad + \quad 2\,H_2O_{(l)}$$

First, it is necessary to find which of the two reactants, $CH_{4(g)}$ or $O_{2(g)}$ is the limiting reagent, then use it to find the masses of $CO_2$ and $H_2O$ produced. To do this, the following steps are carried out.

**Step 2.** Find the number of **moles $CH_4$ in 4.00 g.** To do this, first find its molar mass.

$$M = 1 \times 12.01 \text{ g/mol C} + 4 \times 1.01 \text{ g/mol H} = \textbf{\textit{16.05 g/mol}}$$

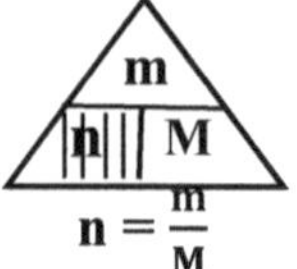

Using the formula, $\mathbf{n = \dfrac{m}{M}}$ we get $n = \dfrac{4.00 \text{ g}}{16.05 \text{ g/mol}}$

$$= \textbf{0.249 mol } \mathbf{CH_4} \text{ (Experimental value)}$$

**Step 3.** Find the number of **moles of $O_2$ in 8.00 g.** To do this, first find its molar mass.

$$M = 2 \times 16.00 \text{ g/mol O} = \textbf{\textit{32.00 g/mol}}$$

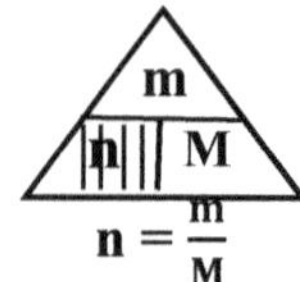

Using the formula, $\mathbf{n = \dfrac{m}{M}}$ we get $n = \dfrac{8.00 \text{ g}}{32.00 \text{ g/mol}}$

$$= \textbf{0.250 mol } \mathbf{O_2} \text{ (Experimental value)}$$

**Step 4.** Find the limiting reagent. To do this, choose anyone of the two reactants and determine how much of the other, in theory, is required to react with it. We choose **0.249 mol $CH_4$** to find how much $O_2$ is required to react with it. In this case, the **0.249 mol $CH_4$** becomes *the given reagent* and $\mathbf{O_2}$ becomes *the required reagent.*

$$\frac{\text{Given}}{\text{Required}} = \frac{1 \text{ mol}}{2 \text{ mol}} \times \frac{0.249 \text{ mol } \mathbf{CH_4}}{\text{X mol } \mathbf{O_2}}$$

$$X = \textbf{0.498 mol } \mathbf{O_2} \text{ (Theoretical value)}$$

$$\textbf{0.250 mol } \mathbf{O_2} \text{ (Experimental value)}$$

Comparison of these two values shows that **less $O_2$**, that is, **(0.250 mol)** was than what was required **(0.498 mol)**. It therefore becomes the *limiting reagent* and **$CH_4$** becomes *the excess reagent.*

**Step 5.** Use the limiting reagent, $O_2$ to calculate the amount of $CO_2$ formed.

$$\frac{\text{Given}}{\text{Required}} = \frac{2 \text{ mol}}{1 \text{ mol}} \times \frac{0.250 \text{ mol } O_2}{\text{X mol } \mathbf{CO_2}}$$

$$\textbf{2 X = 0.250 mol } \mathbf{CO_2} \textbf{ = X = 0.125 mol } \mathbf{CO_2}$$

**Step 6.** Find the mass of **0.125 mol $CO_2$.** To do this, first find its molar mass.

$$M = 1 \times 12.01 \text{ g/mol C} + 2 \times 16.00 \text{ g/mol O} = \textbf{\textit{44.01 g/mol}}$$

Using the formula, $\mathbf{m = n \times M}$ we get

$$\text{Mass} = 0.125 \text{ mol} \times 44.01 \text{ g/mol} = \textbf{5.50 g } \mathbf{CO_2}$$

**Step 7.** Use the limiting reagent, $O_2$ to calculate the amount of $H_2O$ formed.

$$\frac{\text{Given}}{\text{Required}} = \frac{2 \text{ mol}}{2 \text{ mol}} = \frac{0.250 \text{ mol } O_2}{\text{X mol } H_2O}$$

$$\textbf{2 X = 0.500 mol } \mathbf{H_2O} \textbf{ = X = 0.250 mol } \mathbf{H_2O}$$

**Step 8.** Find the mass of 0.250 mol $H_2O$. To do this, first find its molar mass.

$$M = 2 \times 1.01 \text{ g/mol H} + 1 \times 16.00 \text{ g/mol O} = \textbf{\textit{18.02 g/mol}}$$

Using the formula, $\mathbf{m = n \times M}$ we get

$$\text{Mass} = 0.250 \text{ mol} \times 18.02 \text{ g/mol} = \textbf{4.51 g } \mathbf{H_2O}$$

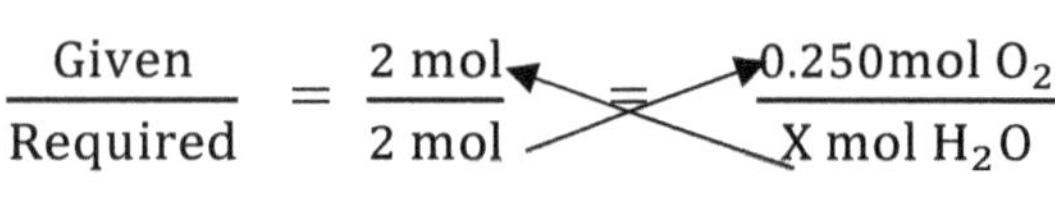

**4.** Calculate the mass of AgCl$_{(s)}$ produced from a reaction between 16.98 g AgNO$_{3(aq)}$ and 10.00 g NaCl$_{(aq)}$. If the actual mass of AgCl$_{(s)}$ produced is 12.10 g, find its percentage yield.

*Provided:*

Mass of AgNO$_3$ = 16.98 g

Mass of NaCl = 10.00 g

Mass of AgCl$_{(s)}$ produced = 12.10 g

*Required:*

The percentage yield of AgCl$_{(s)}$.

*Steps to be carried out:*

**Step 1**. Write a balanced equation for the reaction

$$AgNO_{3(aq)} \quad + \quad NaCl_{(aq)} \quad \rightarrow \quad AgCl_{(s)} \quad + \quad NaNO_{3(l)}$$

First is necessary to find which of the two reactants, AgNO$_3$ or NaCl is the limiting reagent, then use it to find the mass AgCl$_{(s)}$ that should, in theory, be produced. To do this, the following other steps are carried out.

**Step 2.** Find the number of **moles NaCl in the 10.00 g** used. To do this, first find its molar mass.

M = 1 × 22.99 g/mol Na + 1 × 35.45 g/mol Cl = **58.44 g/mol**

Using the formula, $\mathbf{n = \dfrac{m}{M}}$ we get $\mathbf{n} = \dfrac{10.00 \text{ g}}{58.44 \text{ g/mol}}$

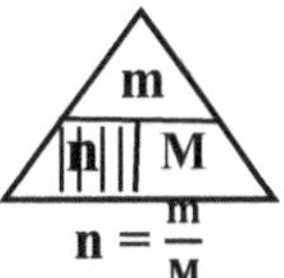


= **0.171mol NaCl** (Experimental value)

**Step 3.** Find the number of moles of **AgNO$_3$ in the 16.98 g** used. To do this, first find its molar mass.

M = 1 × 108.87 g/mol Ag + 1 × 14.01 g/mol N + 3 × 16.00 g/mol O = **170.88 g/mol**

Using the formula, $\mathbf{n = \dfrac{m}{M}}$ we get $\mathbf{n} = \dfrac{16.98 \text{ g}}{170.9 \text{ g/mol}}$

= **0.0991 mol AgNO$_3$** (Experimental value)

**Step 4.** Find the limiting reagent. To do this, choose anyone of the two reactants and determine how much of the other, in theory, is required to react with it. We choose **0.171 mol NaCl** to find how much **AgNO$_3$** is required to react with it. In this case, **the 0.171 mol NaCl** becomes the *given reagent* and **AgNO$_3$** becomes the *required reagent*.

$$\frac{\text{Given}}{\text{Required}} = \frac{1 \text{ mol}}{1 \text{ mol}} = \frac{0.171 \text{ mol } \textbf{NaCl}}{X \text{ mol } \textbf{AgNO}_3}$$

X = **0.171 mol AgNO$_3$** (Theoretical value)

**0.0991 mol AgNO$_3$** (Experimental value)

Comparison of these two values, shows that less AgNO$_3$ was used (**0.099 mol**) than was required (**0.171 mol**); it therefore becomes the *limiting reagent* and **NaCl** becomes the *excess reagent*.

**Step 5.** Use the limiting reagent, **AgNO$_3$** to calculate the amount of **AgCl$_{(s)}$** that should be formed.

$$\frac{\text{Given}}{\text{Required}} = \frac{1 \text{ mol}}{1 \text{ mol}} = \frac{0.0991 \text{ mol AgNO}_3}{X \text{ mol AgCl}}$$

X = **0.0991 mol AgCl**

**Step 6.** Find the mass of **0.0991 mol AgCl.** To do this, first find its molar mass.

$$M = 1 \times 107.87 \text{ g/mol C} + 1 \times 35.45 \text{ g/mol Cl} = \textit{143.32 g/mol}$$

Using the formula, **m = n × M** we get

Mass = 0.0991 mol × 143.32 g/mol = **14.2 g AgCl** (Expected yield)

**Step 7.** Find the percentage yield

$$\text{Pecentage yield} = \frac{\text{Actual yield}}{\text{Exoected yield}} \times 100\,\%$$

$$= \frac{12.10 \text{ g}}{14.2 \text{ g}} \times 100\,\% = \textbf{85.2 \%}$$

## *Problem:*

**5.** Calculate the mass of $BaSO_{4(s)}$ produced from a reaction between 2.61 g $Ba(NO_3)_{2(aq)}$ and 2.00 g $Na_2SO_{4(aq)}$. If the actual mass of $BaSO_{4(s)}$ produced is 2.10 g, find its percentage yield.

*Provided:*

Mass of $Ba(NO_3)_2$ = 2.61 g

Mass of $Na_2SO_4$ = 2.00 g

Mass of $BaSO_{4(s)}$ produced = 2.10 g

*Required:*

Percentage yield of $BaSO_{4(s)}$ produced

*Steps to be carried out:*

**Step 1.** Write a balanced equation for the reaction.

$$Ba(NO_3)_{2(aq)} \quad + \quad Na_2SO_{4(aq)} \quad \rightarrow \quad BaSO_{4(s)} \quad + \quad 2\,NaNO_{3(l)}$$

First is necessary to find which of the two reactants, $Ba(NO_3)_2$ or $Na_2SO_4$, is the limiting reagent, then use it to find the mass of $BaSO_{4(s)}$ that was supposed to be produced. To do this, the following steps are carried out.

**Step 2.** Find the number of **moles of $Na_2SO_4$ in the 2.00 g** used. To do this, first find its molar mass.

$$M = 2 \times 22.99 \text{ g/mol Na} + 1 \times 32.07 \text{ g/mol S} + 4 \times 16.00 \text{ g/mol O} = \textit{142.1 g/mol}$$

Using the formula, $n = \dfrac{m}{M}$ we get $n = \dfrac{2.00 \text{ g}}{142.1 \text{ g/mol}}$

$$= \textbf{0.0140 mol } Na_2SO_4 \text{ (Experimental value)}$$

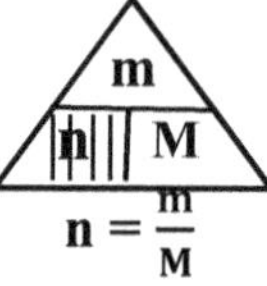


**Step 3.** Find the number of **moles of $Ba(NO_3)_2$ in the 2.61 g** used. To do this, first find its molar mass.

$$M = 1 \times 137.33 \text{ g/mol Ba} + 2 \times 14.01 \text{ g/mol N} + 6 \times 16.00 \text{ g/mol O} = \textit{261.35 g/mol}$$

Using the formula, $n = \dfrac{m}{M}$ we get $n = \dfrac{2.61\,g}{261.35\,g/mol}$

$$= \textbf{0.0100 mol } Ba(NO_3)_2 \text{ (Experimental value)}$$

**Step 4.** Find the limiting reagent. To do this, choose anyone of the two reactants and determine how much of the other, in theory, is required to react with it. We chose **0.0100 mole of Ba(NO₃)₂** to find how much **Na₂SO₄** is required to react with it. In this case, the **0.01000 mole of Ba(NO₃)₂** becomes the *given reagent* and **Na₂SO₄** becomes the *required reagent.*

$$\frac{\text{Given}}{\text{Required}} = \frac{1 \text{ mol}}{1 \text{ mol}} \quad\rightleftarrows\quad \frac{0.0100 \text{ mol Ba(NO}_3)_2}{\text{X mol Na}_2\text{SO}_4}$$

$$X = 0.0100 \text{ mol Na}_2\text{SO}_4 \text{ (Theoretical value)}$$

$$0.0140 \text{ mol Na}_2\text{SO}_4 \text{ (Experimental value)}$$

Comparison of these two values shows that more **Na₂SO₄,** that is, **(0.014 mol)** was used than the **(0.010 mol)** that was required; it therefore becomes the *excess reagent* and that **Ba(NO₃)₂** *the limiting reagent.*

$$\text{Excess amount of Na}_2\text{SO}_4 = 0.0140 \text{ mol} - 0.0100 \text{ mol} = \textbf{0.00400 mol}$$

**Step 4.** Use the limiting reagent, Ba(NO₃)₂ to calculate the amount of BaSO₄(s) formed.

$$\frac{\text{Given}}{\text{Required}} = \frac{1 \text{ mol}}{1 \text{ mol}} \quad\rightleftarrows\quad \frac{0.0100 \text{ Ba(NO}_3)_2}{\text{X mol BaSO}_4}$$

$$X = \textbf{0.0100 mol BaSO}_4{}_{(s)}$$

**Step 5.** Find the mass of **0.0100 mol BaSO₄(s).** To do this, first find its molar mass.

$$M = 1 \times 137.33 \text{ g/mol Ba} + 1 \times 32.07 \text{ g/mol S} + 4 \times 16.00 \text{ g/mol O} = \textit{233.4 g/mol}$$

Using the formula, **m = n × M** we get

$$\text{Mass} = 0.0100 \text{ mol} \times 233.4 \text{ g/mol} = \textbf{2.33 g BaSO}_4{}_{(s)} \text{ (Expected yield)}$$

**Step 6.** Find the percentage yield.

$$\text{Pecentage yield} = \frac{\text{Actual yield}}{\text{Expected yield}} \times 100 \%$$

$$= \frac{2.10 \text{ g}}{2.33 \text{ g}} \times 100 \% = \textbf{90.1 \%}$$

## *Problem:*

6. What mass of PbI₂(s) is produced from the reaction between 4.00 g NaI(aq) and 15.00 g Pb(NO₃)₂(aq)? If the actual mass of PbI₂(s) produced is 5.21 g, find its percentage yield.

*Provided:*

Mass of Pb(NO₃)₂ = 15.00 g

Mass of NaI  = 4.00 g

Mass of PbI₂(s) produced = 5.21 g

*Required:*

The percentage yield of PbI₂(s)

*Steps to be carried out:*

**Step 1.** Write a balanced equation for the reaction

$$\text{Pb(NO}_3)_{2(aq)} + 2 \text{ NaI}_{(aq)} \rightarrow 2 \text{ NaNO}_{3(aq)} + \text{PbI}_{2(S)}$$

First, it is necessary to find which of the two reactants, Ba(NO₃)₂ or Na₂SO₄ is the limiting reagent, then use it to find the mass of BaSO₄(s). To do this, the following steps are carried out.

**Step 2**. Find the number of **moles of NaI in the 4.00 g** used. To do this, first find its molar mass.

$$M = 1 \times 22.99 \text{ g/mol Na} + 1 \times 126.90 \text{ g/mol} = \textbf{\textit{149.9 g/mol}}$$

Using the formula, $\mathbf{n = \dfrac{m}{M}}$ we get $\mathbf{n} = \dfrac{4.00 \text{ g}}{149.9 \text{ g/mol}}$

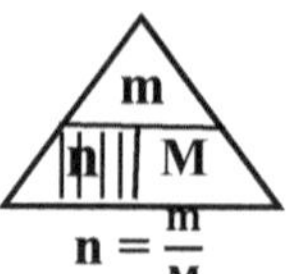

$$= \textbf{0.0266 mol NaI}_{(aq)} \text{ (Experimental value)}$$

**Step 3**. Find the number of **moles of Pb(NO₃)₂ in the 15.00 g** used. To do this, first find its molar mass.

$$M = 1 \times 207.20 \text{ g/mol Pb} + 2 \times 14.01 \text{ g/mol N} + 6 \times 16.00 \text{ g/mol O} = \textbf{\textit{331.22 g/mol}}$$

Using the formula, $\mathbf{n = \dfrac{m}{M}}$ we get $\mathbf{n} = \dfrac{15.00 \text{ g}}{331.22 \text{ g/mol}}$

$$= \textbf{0.0452 mol Pb(NO}_3\textbf{)}_2 \text{ (Experimental value)}$$

**Step 3**. Find the limiting reagent. To do this, choose any one of the two reactants and determine how much of the other, in theory, is required to react with it. We chose **0.0452 mol Pb(NO₃)₂** to find how much **NaI** is required to react with it. In this case, the **0.0452 mol Pb(NO₃)₂** becomes *the given reagent* and **NaI** *the required reagent*.

$$\frac{\text{Given}}{\text{Required}} = \frac{1 \text{ mol}}{2 \text{ mol}} \quad\times\quad \frac{0.0452 \text{ mol } \mathbf{Pb(NO_3)_2}}{\text{X mol } \mathbf{NaI}}$$

$$\text{X} = \textbf{0.0452 mol NaI} \text{ (Theoretical value)}$$

$$\textbf{0.0266 mol NaI} \text{ (Experimental value)}$$

Comparison of these two values shows that **less NaI, that is, (0.0266 mol)** was used than the **(0.0452 mol)** that was requried. It therefore becomes *the limiting reagent* and **Pb(NO₃)₂** the *excess reagent.*

**Step 4**. Use the limiting reagent, NaI to calculate the amount of PbI₂(s) formed.

$$\frac{\text{Given}}{\text{Required}} = \frac{2 \text{ mol}}{1 \text{ mol}} \quad\times\quad \frac{0.0266 \text{ mol NaI}}{\text{X mol PbI}_2}$$

$$\textbf{2 X} = \textbf{0.0266 mol PbI}_2 = \textbf{X} = \textbf{0.0133 mol PbI}_2$$

**Step 5**. Find the mass of **0.0133 mol PbI₂(s)**. To do this, first find its molar mass.

$$M = 1 \times 207.20 \text{ g/mol Pb} + 2 \times 126.9 \text{ g/mol I} = \textbf{\textit{461.0 g/mol}}$$

Using the formula, $\mathbf{m = n \times M}$ we get

Mass $= 0.0133 \text{ mol} \times 461.0 \text{ g/mol} = \textbf{6.13 g PbI}_2$ (Expected yield)

**Step 6**. Find the percentage yield

$$\text{Pecentage yield} = \frac{\text{Actual yield}}{\text{Exoected yield}} \times 100 \%$$

$$= \frac{5.21 \text{g}}{6.13 \text{ g}} \times 100 \% = \textbf{85.0\%}$$

*Problem:*

7 . Aspirin is made from the reaction between salicylic acid and acetic anhydride according to the following equation:

| $C_7H_6O_3$ | + | $C_4H_6O_3$ | $\longrightarrow$ | $C_9H_8O_4$ | + | $C_2H_4O_2$ |
|---|---|---|---|---|---|---|
| salicylic acid | | acetic anhydride | | aspirin | | acetic acid |

a) What mass of aspirin can be produced if 10.0 g portions of each reactant were reacted in an experiment?

*Provided:*

Mass of $C_7H_6O_3 = 10.0$ g

Mass of $C_4H_6O_3 = 10.0$ g

*Required:*

The mass of aspirin ($C_9H_8O_4$) produced

*Steps to be carried out:*

**Step 1.** Write a balanced equation for the reaction.

$$C_7H_6O_3 \quad + \quad C_4H_6O_3 \quad \longrightarrow \quad C_9H_8O_4 \quad + \quad C_2H_4O_2$$

salicylic acid    acetic anhydride                aspirin      acetic acid

First, it is necessary to find which of the two reactants, $C_7H_6O_3$ and $C_4H_6O_3$, is the limiting reagent, then use it to find the mass of $C_9H_8O_4$ produced. To do this, the following steps are carried out.

**Step 2.** Find the number of **moles of $C_7H_6O_3$ in the 10.0 g** used. To do this, first find its molar mass.

$$M = 7 \times 12.01 \text{ g/mol C} + 6 \times 1.01 \text{ g/mol H} + 3 \times 16.00 \text{ g/mol O} = \textit{138.1 g/mol}$$

Using the formula, $n = \dfrac{m}{M}$ we get $n = \dfrac{10.00\ g}{138.1\ g/mol}$

$$= \textbf{0.0724 mol } \mathbf{C_7H_6O_3}\ (\text{Experimental value})$$

**Step 3.** Find the number of moles of $\mathbf{C_4H_6O_3}$ **in 10.0 g** used. To do this, first find its molar mass.

$$M = 4 \times 12.01 \text{ g/mol C} + 6 \times 1.01 \text{ g/mol H} + 3 \times 16.00 \text{ g/mol O} = \textit{102.1 g/mol}$$

Using the formula, $\mathbf{n} = \dfrac{\mathbf{m}}{\mathbf{M}}$ we get $\mathbf{n} = \dfrac{10.0 \text{ g}}{102.1 \text{ g/mol}}$

$$= \textbf{0.0979 mol } \mathbf{C_4H_6O_3}\ (\text{Experimental value})$$

**Step 4.** Find the limiting reagent. To do this, choose any one of the two reactants and determine how much of the other, in theory, is required to react with it. We chose **0.0724 mol $C_7H_6O_3$**, to find how much **$C_4H_6O_3$** is required to react with it. In this case, **0.0724 mol $C_7H_6O_3$**, becomes *the given reagent* and **$C_4H_6O_3$** becomes *the required reagent.*

$$\frac{\text{Given}}{\text{Required}} = \frac{1 \text{ mol}}{1 \text{ mol}} \diagdown \frac{0.0724 \text{ mol } C_7H_6O_3}{X \text{ mol } C_4H_6O_3}$$

$$X = \textbf{0.0724 mol } \mathbf{C_4H_6O_3}\ (\text{Theoretical value})$$

$$\textbf{0.0979 mol } \mathbf{C_4H_6O_3}\ (\text{Experimental value})$$

Comparison of these two values shows that more $C_4H_6O_3$ that is, **(0.0979 mol)** was used than what was required; it therefore becomes the *excess reagent* and $\mathbf{C_7H_6O_3}$ *the limiting reagent.*

The amount of excess $C_4H_6O_3 = 0.0979$ mol $- 0.0724$ mol $= \textbf{0.025 mol}$

**Step 5.** Use the limiting reagent, $C_7H_6O_3$ to calculate the amount of $C_9H_8O_4$ formed.

$$\frac{\text{Given}}{\text{Required}} = \frac{1 \text{ mol}}{1 \text{ mol}} \diagdown \frac{0.0724 \text{ mol } C_7H_6O_3}{X \text{ mol } C_9H_8O_4}$$

$$X = \textbf{0.0724 mol } \mathbf{C_9H_8O_4}$$

**Step 5.** Find the mass of **0.072 4 mol $C_9H_8O_4$.** To do this, first find its molar mass.

M = 9 × 12.01 g/mol C + 8 × 1.01 g/mol H + 4 × 16.00 g/mol O = *180.2 g/mol*

Using the formula, **m = n × M** we get

Mass = 0.0724 mol × 180.2 g/mol = **13.0 g $C_9H_8O_4$** (aspirin)

## *Problem:*

**8.** One form of soap is made by the reaction between stearic acid and sodium hydroxide according to the following chemical equation:

**Stearic acid**            **Soap**

$NaOH_{(s)}$ + $C_{17}H_{35}COOH$ → $C_{17}H_{35}COONa_{(s)}$ + $H_2O_{(l)}$

a) If 4.00 g of $NaOH_{(s)}$ reacts with 3.00 g of $C_{17}H_{35}COOH$, calculate what mass of soap would be formed.

b) What reagent is in excess?

c) What side effects would this soap have on people's skin?

*Provided:*

Mass of $NaOH_{(s)}$ = 4.00 g

Mass of $C_{17}H_{35}COOH$ = 3.00 g

*Required:*

The mass of soap produced

*Steps to be carried out:*

**Step 1.** Write a balanced equation for the reaction.

$NaOH_{(s)}$ + $C_{17}H_{35}COOH$ → $C_{17}H_{35}COONa_{(s)}$ + $H_2O_{(l)}$

First, it is necessary to find which of the two reactants, NaOH or $C_{17}H_{35}COOH$ is the limiting reagent, then use it to find the mass of $C_{17}H_{35}COONa_{(s)}$. To do this, the following steps are carried out.

**Step 2.** Find the number of moles of NaOH. To do this, first find its molar mass.

M = 1 × 22.99 g/mol Na + 1 × 1.01 g/mol H + 1 × 16.00 g/mol O = *40.0 g/mol*

Using the formula, **n = $\frac{m}{M}$** we get **n =** $\dfrac{4.00 \text{ g}}{40.0 \text{ g/mol}}$

= **0.100 mol NaOH** (Experimental value)

**Step 3.** Find the number of moles of $C_{17}H_{35}COOH$. To do this, first find its molar mass.

M = 18 × 12.01 g/mol C + 36 × 1.01 g/mol H + 2 × 16.00 g/mol O = *284.5 g/mol*

Using the formula, **n = $\frac{m}{M}$** we get **n =** $\dfrac{3.00 \text{ g}}{284.5 \text{ g/mol}}$

= **0.0104 mol $C_{17}H_{35}COOH$** (Steric acid) (Experimental value)

**Step 4.** Find the limiting reagent. To do this, choose anyone of the two reactants and determine how much of the other, in theory, is required to react with it. We choose **0.100 mol NaOH** to find how much $C_{17}H_{35}COOH$ is required to react with it. In this case, the **0.100 mol NaOH** becomes *the given reagent* and $C_{17}H_{35}COOH$ becomes *the required reagent*.

$$\frac{\text{Given}}{\text{Required}} = \frac{1\ \text{mol}}{1\ \text{mol}} \quad = \quad \frac{0.100\ \text{mol NaOH}}{X\ \text{mol } C_{17}H_{35}\ COOH}$$

$$X = \textbf{0.100 mol } \mathbf{C_{17}H_{35}COOH}\ \text{(Theoretical value)}$$

$$\textbf{0.0104 mol } \mathbf{C_{17}H_{35}COOH}\ \text{(Experimental value)}$$

Comparison of these two values shows that less $C_{17}H_{35}COOH$, that is, **(0.0104 mol)** was used than what was required **(0.100 mol)**; it therefore becomes *the limiting reagent* and **NaOH** *the excess reagent.*

**Step 5.** Use the limiting reagent to calculate the amount of soap ($C_{17}H_{35}COONa$) formed.

$$\frac{\text{Given}}{\text{Required}} = \frac{1\ \text{mol}}{1\ \text{mol}} \quad \frac{0.0104\ \text{mol } C_{17}H_{35}COOH}{X\ \text{mol } C_{17}H_{35}COONa}$$

$$X = \textbf{0.0104 mol } \mathbf{C_{17}H_{35}\,COONa}$$

**Step 6.** Find the mass of **0.0104 mol $C_{17}H_{35}$ COONa.** To do this, first find its molar mass.

M = 18 ×12.01 g/mol C+35 ×1.01 g/mol H+2 ×16.00 g/mol O+1 ×22.99 g/mol Na = **306.5 g/mol**

Using the formula, $\mathbf{m = n \times M}$ we get

Mass = 0.0104 mol × 306.5 g/mol = **3.18 g $\mathbf{C_{17}H_{35}}$ COONa** (Soap)

Because NaOH which is caustic in nature, an excess of it may cause irritation to the skin if it is left in the soap.

### *Problem:*

**9.** In an attempt to test the gravimetric stoichiometric method, a student mixed 150.0 mL of 0.15 mol/L $Na_2CO_{3(aq)}$ solution with 200.0 mL of 0.12 mol/L $BaCl_{2(aq)}$ solution to a white precipitate of $BaCO_{3(s)}$. The precipitate was recovered from the mixture using the filtration method. The precipitate, after being dried, had a mass of 4.42 g. Find the percent yield. Is the gravimetric stoichiometric valid?
* This problem can't be solved by the students at this stage of the course since the concept of aqueous solution has not been taught. But it will be solved for later reference.

*Provided:*
Mass of $BaCO_{3(s)}$ obtained = 4.42 g

Volume of $Na_2CO_{3(aq)}$ =150.0 mL

Concentration of $Na_2CO_{3(aq)}$ = 0.15 mol/L

Volume of $BaCl_{2(aq)}$ = 200.0 mL

Concentration of $BaCl_{2(aq)}$ = 0.12 mol/L

*Required:*

The percentage yield of $BaCO_{3(s)}$.

*Steps to be carried out:*

**Step 1.** Write a balanced equation for the reaction

$$Na_2CO_{3(aq)} \quad + \quad BaCl_{2\,(aq)} \quad \rightarrow \quad BaCO_{3(s)} \quad + \quad 2\,NaCl_{(aq)}$$

First, it is necessary to find which of the two reactants, $Na_2CO_{3(aq)}$ or $BaCl_{2(aq)}$ is the limiting reagent, then use it to find the mass of $BaCO_{3(s)}$. To do this, the following steps are carried out.

**Step 2.** Find the number of moles of $Na_2CO_{3(aq)}$. Since this is an aqueous solution, the formula, $n = C \times V$ is used to find number of moles. In this equation, C represents the concentration of the solution and V, its volume in litres. Since the volume is given in mL it must be changed to litres. This is done by dividing by 1000 mL/L, that is, 150 mL $\div$ 1000 mL/L = 0.150 L

$$\mathbf{n = C \times V} = 0.15 \text{ mol/L } \times 0.150 \text{ L } = \mathbf{0.022 \text{ mol } Na_2CO_{3(aq)}}$$

**Step 3.** Find the number of moles of $BaCl_{2(aq)}$. Doing the same as in step **2**, we get for the volume:

$$V = 200.0 \text{ mL} \div 1000 \text{ mL/L} = 0.200 \text{ L}$$

$$\mathbf{n = C \times V} = 0.12 \text{ mol/L } \times 0.200 \text{ L } = \mathbf{0.024 \text{ mol } BaCl_{2(aq)}.}$$

**Step 4.** Find the limiting reagent. To do this, choose anyone of the two reactants and determine how much of the other, in theory, is required to react with it. We choose **0.024 mol $BaCl_{2(aq)}$** to find how much **$Na_2CO_{3(aq)}$** is required to react with it. In this case, the **0.024 mol $BaCl_{2(aq)}$** becomes *the given reagent* and **$Na_2CO_{3(aq)}$** becomes *the required reagent*.

$$\frac{\text{Given}}{\text{Required}} = \frac{1 \text{ mol}}{1 \text{ mol}} \quad = \quad \frac{0.024 \text{ mol } BaCl_{2(aq)}}{X \text{ mol } Na_2CO_{3(aq)}}$$

$$\mathbf{X = 0.024 \text{ mol } Na_2CO_{3(aq)}} \text{ (Theoretical value)}$$

$$\mathbf{0.022 \text{ mol } Na_2CO_{3(aq)}} \text{ (Experimental value)}$$

Comparison of these two values shows that less $Na_2CO_3$ was used, that is **(0.022 mol)** was used than what was required **(0.024 mol)**; it therefore becomes *the limiting reagent* and the **$BaCl_{2(aq)}$** *excess reagent.*

**Step 5.** Use the limiting reagent, **$Na_2CO_{3(aq)}$** calculate the amount of $BaCO_{3(s)}$ formed.

$$\frac{\text{Given}}{\text{Required}} = \frac{1 \text{ mol}}{1 \text{ mol}} \quad \frac{0.022 \text{ mol } Na_2CO_{3(aq)}}{X \text{ mol } BaCO_{3(s)}}$$

$$\mathbf{X = 0.022 \text{ mol } BaCO_{3(s)}}$$

**Step 6.** Find the mass of 0.023 mol $BaCO_{3(s)}$. To do this, first find its molar mass.

$$M = 1 \times 137.33 \text{ g/mol Ba} + 1 \times 12.01 \text{ g/mol C} + 3 \times 16.00 \text{ g/mol} = \mathit{197.3 \text{ g/mol}}$$

Using the formula, $\mathbf{m = n \times M}$ we get

$$\text{Mass} = 0.022 \text{ mol} \times 197.34 \text{ g/mol} = \mathbf{4.34 \text{ g } BaCO_{3(s)}} \text{ (Expected yield)}$$

$$m = n \times M$$

**Step 7.** Find the percentage yield.

$$\text{Pecentage yield} = \frac{\text{Actual yield}}{\text{Exoected yield}} \times 100 \text{ \%}$$

$$= \frac{4.34 \text{g}}{4.54 \text{ g}} \times 100 \text{ \% } = \mathbf{95.6 \text{ \%}}$$

Based on the results, we can conclude that the gravimetric stoichiometric method is indeed valid.

# Chapter Review: Quantities in chemical reactions (9-11) solutions

## Matching

*Match each term in the table below with the correct statements that follow*

| | | | |
|---|---|---|---|
| A | Isotopes | G | Relative atomic mass |
| B | Molar mass | H | Molecular formula |
| C | One mole | I | Avogadro's number |
| D | Empirical formula | J | Excess reagent |
| E | Formula unit | K | Stoichiometry |
| F | Limiting reagent | L | Hydrates |

1. The study of the relationship between the quantities of reactants used and products formed in chemical reactions.

2. This represents the smallest whole number ratio of the atoms of elements present in a compound.

3. Atoms of the same element having the same atomic number but different mass number

4. This reagent is left over after a chemical reaction.

5. These compounds incorporate molecules of water in their crystals.

6. The reactant that is completely consumed in a chemical reaction.

7. This represents the actual number of atoms of each element present in one molecule of a substance.

8. This is the average mass of all the naturally occurring isotopes in an element.

9. The amount of a substance that contains the same number of entities as there are atoms in 12 g of carbon-12.

10. The smallest repeating unit in any ionic compound.

11. This number represents $6.02 \times 10^{23}$ atoms, ions or molecules or other entities.

12. The mass of one mole of atoms, molecules or formula units.

## True or False

**Read each of the following statements and then decide if it is *True* or *False*.**

| | |
|---|---|
| 1. | The mass of one proton is approximately equal to the mass of one neutron. |
| 2. | The empirical formula of a compound represents the smallest whole number ratio of the atoms of elements present in a compound. |
| 3. | When calculating the molar mass of a hydrated compound, the mass of water is not included. |
| 4. | $CH_2$ is the empirical formula for benzene ($C_6H_6$). |
| 5. | The mass of one formula unit of zinc chloride is the sum of the masses of two zinc ions and one chloride ion. |
| 6. | The molar mass for ions is the same as that for mole of atoms for the same element. |
| 8. | Two different compounds cannot have identical percentage compositions. |
| 9. | The amount of the limiting reagent determines the amount of product(s) formed in a chemical reaction. |
| 10. | If 1.2 g of magnesium oxide was produced in a reaction for which the theoretical yield was 1.4 g, then the percentage yield is 85.71%. |
| 11. | It is of no consequences if excess reagents are left in medicines and pharma products. |

## Multiple Choice

**Choose the letter that best answers the questions.**

1. The atomic notation for carbon is $_6^{12}C$ . The nucleus for this atom will have

| | | | | |
|---|---|---|---|---|
| *a* | 12 neutrons and 6 protons | *d* | 12 neutrons and 12 protons |
| *b* | 6 neutrons and 6 protons | *e* | none of the above |
| *c* | 12 protons and 6 neutrons | | |

2. Magnesium has three naturally occurring isotopes, as follows: 78.99 % $^{24}$Mg, 10.00 % $^{25}$Mg and 11.01 % $^{26}$Mg. The average atomic mass of Magnesium is

| | | | | |
|---|---|---|---|---|
| *a* | 28.1 u | *d* | 25.0 u |
| *b* | 243.1 u | *e* | 24.0 u |
| *c* | 24.31 u | | |

3.

The mass of one hydrogen atom is approximately equal to

| | | | |
|---|---|---|---|
| a | 1.0 g | d | 6.023 x10$^{23}$ u |
| b | 2.0 g | e | 1.0 u |
| c | 2.0 u | | |

4.

The mass of one $^{12}_{6}C$ atom approximately equals the mass of

| | | | |
|---|---|---|---|
| a | 1 $^{1}_{1}H$ atom | d | 1 $^{13}_{6}C$ atom |
| b | 12 $^{1}_{1}H$ atoms | e | 6 $^{1}_{1}H$ atoms |
| c | 1 $^{14}_{7}N$ atom | | |

5.

The molar mass of magnesium sulphate, $MgSO_4 \cdot 7H_2O$ (Epson salt) is

| | | | |
|---|---|---|---|
| a | 246.52 g/mol | d | 159.5 g/mol |
| b | 120.38 g/mol | e | none of the above |
| c | 126.14 g/mol | | |

6.

The ratio of carbon to oxygen, by mass, in carbon dioxide is

| | | | |
|---|---|---|---|
| a | 2:1 | d | 1:3 |
| b | 3:8 | e | 12:16 |
| c | 12:1 | | |

7.

The number of **moles** of molecules in 8.0 g of oxygen, (molar mass = 32.0 g/mol) is

| | | | |
|---|---|---|---|
| a | 0.50 mol | d | 32 mol |
| b | 0.25 mol | e | 1.0 mol |
| c | 80 mol | | |

8.

The mass of 0.250 moles of copper(II) sulphate pentahydrate, $CuSO_4 \cdot 5H_2O$, is

| | | | |
|---|---|---|---|
| a | 62.4 g | d | 39.87 g |
| b | 124.7 g | e | 22.50 g |
| c | 249.4 g | | |

9.

The percentage composition of nitrogen, by mass, in the compound, $NH_4OH$ is

| | | | |
|---|---|---|---|
| a | 47.0 % | d | 20.0 % |
| b | 34.2 % | e | 18.5 % |
| c | 39.96 % | | |

10. The number of **molecules** in 16.0 g of oxygen is

| | | | |
|---|---|---|---|
| a | $6.02 \times 10^{23}$ | d | $3.01 \times 10^{23}$ |
| b | $1.50 \times 10^{23}$ | e | $1.5 \times 10^{22}$ |
| c | $1.20 \times 10^{24}$ | | |

11. The fertilizer, ammonium nitrate has the formula, $(NH_4)_3PO_4$. The percentage of phosphorus, by mass, in this compound is

| | | | |
|---|---|---|---|
| a | 8.10% | d | 42.9% |
| b | 25.5% | e | 35.5% |
| c | 20.7% | | |

12. A 4.03 g sample of a compound contains 2.43 g of magnesium and 1.60 g of oxygen. The empirical formula of the compound is

| | | | |
|---|---|---|---|
| a | $Mg_2O$ | d | $Mg_2O_2$ |
| b | $MgO_2$ | e | $MgO_2$ |
| c | $MgO$ | | |

13. A compound is found to consist of 39.99 % carbon, 6.73 % hydrogen, and 53.28 % oxygen, by mass. If the molar mass of the compound is 180.18 g/mol, it molecular formula is

| | | | |
|---|---|---|---|
| a | $CH_2O$ | d | $C_6H_{12}O_6$ |
| b | $C_3H_6O_3$ | e | $CHO$ |
| c | $C_2HO$ | | |

14. Oxygen gas can be produced by the decomposition of potassium chlorate, as shown by the following equation: $2\ KClO_{3(s)} \longrightarrow 2\ KCl + 3\ O_2$. If 4 moles of $KClO_{3(s)}$ are decomposed, the mass of oxygen produced will be

| | | | |
|---|---|---|---|
| a | 96.0 g | d | 16.0 g |
| b | 192 g | e | 64.0 g |
| c | 48.0g | | |

15. Sulphur trioxide is produced from a reaction as shown in the equation:

$2\ SO_{2(g)} + O_{2(g)} \longrightarrow 2\ SO_{3(g)}$ The number of moles of oxygen required to produce 8 moles of sulphur trioxide is

| | | | |
|---|---|---|---|
| a | 16 | d | 8 |
| b | 4 | e | 12 |
| c | 32 | | |

16.

In an experiment, substances X and Y react according the following equation:

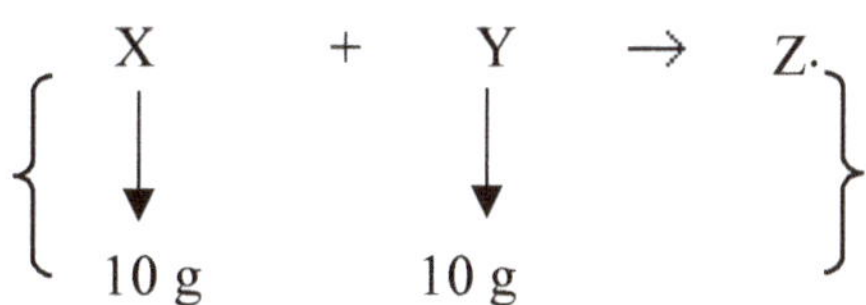

The mass of Z formed is

| | | | | |
|---|---|---|---|---|
| a | 20 g | d | 15 g |
| b | 10.0 g | e | not possible to be calculated from the information given. |
| c | 5.0 g | | |

17.

Which of the following statements is **not true** about the **limiting reagent** in a chemical reaction?

| | |
|---|---|
| a | It is the reactant that is not consumed completely |
| b | It is the reactant that is consumed completely |
| c | It is the reactant that determines how much of the other reactant(s) is consumed |
| d | It is the reactant that determines how much product is formed |
| e | b, c and d |

18.

Ammonia is produced from a reaction as shown in the equation:

$3\ H_{2(g)} + N_{2(g)} \longrightarrow 2\ NH_{3(g)}$ The number of **moles** of hydrogen required to produce 8.0 moles of ammonia if the percentage yield is 25 % is

| | | | | |
|---|---|---|---|---|
| a | 12 | d | 8.0 |
| b | 48 | e | 4.0 |
| c | 32 | | |

19.

Hydrogen peroxide decomposes according to the following chemical equation:

$$2\ H_2O_{2(l)} \longrightarrow 2\ H_2O_{(l)} + O_{2(g)}$$

What mass of $H_2O_{2(l)}$ must be decomposed to produce 16.00 g of $O_{2(g)}$?

| | | | | |
|---|---|---|---|---|
| a | 68.04 g | d | 17.01 g |
| b | 50.0 g | e | 34.02 g |
| c | 32.0 g | | |

20.

Magnesium reacts with hydrochloric according to the following chemical equation:
$$Mg_{(s)} + 2\ HCl_{(aq)} \longrightarrow MgCl_{2(aq)} + H_{2(g)}$$
What mass of $Mg_{(s)}$ is required to react completely with 100.0 mL of 1.0 mol/L $HCl_{(aq)}$?

| | | | | |
|---|---|---|---|---|
| a | 2.43 g | d | 243 g |
| b | 24.3 g | e | 0.120 g |
| c | 1.22 g | | |

21.

What mass of calcium hydroxide is required to react completely with 200.0 mL of 0.100 mol/L $HCl_{(aq)}$ according to the following equation:

$$Ca(OH)_{2(s)} + 2\ HCl_{(aq)} \longrightarrow CaCl_{2(aq)} + 2\ H_2O_{(l)}$$

| | | | |
|---|---|---|---|
| *a* | 2.43 g | *d* | 7.41 g |
| *b* | 24.3 g | *e* | 0.120 g |
| *c* | 1.22 g | | |

22.

Determine the mass of copper that is produced when 2.43 g of magnesium reacts with excess copper (11) sulphate solution as shown in the following reaction:

$$Mg_{(s)} + CuSO_{4(aq)} \longrightarrow MgSO_{4(aq)} + Cu_{(s)}$$

| | | | |
|---|---|---|---|
| *a* | 6.36 g | *d* | 3.18 g |
| *b* | 63.55 g | *e* | 2.43 g |
| *c* | 0.630 g | | |

23.

Hydrochloric acid and sodium hydroxide react according to the following equation:

$$HCl_{(aq)} + NaOH_{(aq)} \longrightarrow NaCl_{(aq)} + H_2O_{(l)}$$

If 200.0 mL of 0.20 mol/L $HCl_{(aq)}$ mixes with 150.0 mL of 0.30 mol/L $NaOH_{(aq)}$. The limiting reagent is

| | | | |
|---|---|---|---|
| *a* | $NaOH_{(aq)}$ | *d* | $H_2O_{(l)}$ |
| *b* | $NaCl_{(aq)}$ | *e* | None of the above |
| *c* | $HCl_{(aq)}$ | | |

24.

Nitric acid reacts with sodium carbonate according to the following equation:

$$2\ HNO_{3(aq)} + Na_2CO_{3(s)} \longrightarrow 2\ NaNO_{3(aq)} + H_2O_{(l)} + CO_{2(g)}$$

The mass of $CO_{2(g)}$ formed when 200.0 mL of 0.10.0 mol/L $HNO_{3(aq)}$ reacts with 2.00 g of $Na_2CO_{3(s)}$ is

| | | | |
|---|---|---|---|
| *a* | 2.00 g | *d* | 0.830 g |
| *b* | 8.30 g | *e* | 0.440 g |
| *c* | 4.40 g | | |

25.

The mass of copper produced is 12.01 g when a mass of 4.86 g of magnesium reacts with excess copper (11) sulphate solution as shown in the following reaction:

$$Mg_{(s)} + CuSO_{4(aq)} \longrightarrow MgSO_{4(aq)} + Cu_{(s)}$$

The percentage yield of copper is

| | | | |
|---|---|---|---|
| *a* | 4.9 % | *d* | 71.0 % |
| *b* | 94.4 % | *e* | 58.3 % |
| *c* | 60.2 % | | |

26.

Magnesium hydroxide, $Mg(OH)_2$, reacts with hydrochloric acid, $HCl_{(aq)}$ according to the equation: $Mg(OH)_{2(s)} + 2\ HCl_{(aq)} \longrightarrow MgCl_{2(aq)} + 2\ H_2O_{(l)}$. If an antacid tablet has $Mg(OH)_{2(s)}$ as its main ingredient, what mass of it must be present to neutralize 200.0 mL of stomach fluid that has $HCl_{(aq)}$ of concentration 0.10 mol/L?

| | | | | |
|---|---|---|---|---|
| *a* | 0.58 g | | *d* | 71 g |
| *b* | 1.16 g | | *e* | 58.3 g |
| *c* | 5.83 g | | | |

27.

Aqueous sodium carbonate reacts with aqueous barium chloride according to the following equation: $Na_2CO_{3(aq)} + BaCl_{2(aq)} \longrightarrow BaCO_{3(s)} + 2\ NaCl_{(aq)}$.

When 5.30 g of $Na_2CO_{3(aq}$ is mixed with 15.4 g of $BaCl_{2(aq}$ in an experiment, 9.10 g of $BaCO_{3(s)}$ was obtained after the mixture was filtered and the precipitate was dried. The percent yield of $BaCO_{3(s)}$ was

| | | | | |
|---|---|---|---|---|
| *a* | 95.0% | | *d* | 75.5% |
| *b* | 100% | | *e* | 80% |
| *c* | 92.2% | | | |

28.

Ethanol burns with oxygen according to the following equation:

$$C_2H_5OH_{(l)} + 3\ O_{2g)} \longrightarrow 2\ CO_{2(g)} + 3\ H_2O_{(g)}$$

When 3.45 g of $C_2H_5OH_{(l)}$ was burnt in 7.80 g of $O_{2(g)}$, the mass of $CO_{2(g)}$ produced was found to be 6.32 g. What was the percentage yield of $CO_{2(g)}$?

| | | | | |
|---|---|---|---|---|
| *a* | 95.8% | | *d* | 90% |
| *b* | 98% | | *e* | 85 % |
| *c* | 92% | | | |

29.

Aqueous potassium nitrate reacts with aqueous silver nitrate according to the following equation: $KI_{(aq)} + AgNO_{3(aq)} \longrightarrow KNO_{3(aq)} + AgI_{(s)}$

If 2.16 g of $AgI_{(s)}$ was obtained and this represented 92.0 % yield, what possible mass combination of $KI_{(aq)}$ and $AgNO_{3(aq)}$ can produce this amount of precipitate?

| | | | | |
|---|---|---|---|---|
| *a* | 1.06 g $KI_{(aq)}$ : 2.49 $AgNO_{3(aq}$ | | *d* | all of the above |
| *b* | 1.71 g $KI_{(aq)}$ : 2.49 $AgNO_{3(aq}$ | | *e* | none of the above |
| *c* | 1.06 g $KI_{(aq)}$ : 2.49 $AgNO_{3(aq}$ | | | |

# Chapter Review: Quantities in Chemical reactions (9-11) solutions

## Matching :

| | | | | | | | |
|---|---|---|---|---|---|---|---|
| 1.) K | 2.) D | 3.) A | 4.) J | 5.) L | 6.) F | 7.) H | 8.) G |
| 9.) I | 10.) E | 11.) C | 12.) B | | | | |

## True/False

| | | | | | | | |
|---|---|---|---|---|---|---|---|
| 1.) T | 2.) T | 3.) T | 4.) F | 5.) F | 6.) T | 7.) T | 8.) F |
| 9.) T | 10.) T | 11.) F | | | | | |

## Multiple Choice

| | | | | | | | |
|---|---|---|---|---|---|---|---|
| 1.) B | 2.) C | 3.) E | 4.) C | 5.) A | 6.) B | 7.) B | 8.) A |
| 9.) C | 10.) D | 11.) C | 12.) C | 13.) D | 14.) B | 15.) B | 16.) E |
| 17.) B | 18.) B | 19.) E | 20.) C | 21.) D | 22.) A | 23.) C | 24.) E |
| 25.) B | 26.) A | 27.) C | 28.) A | 29.) D | | | |

# Unit 3: Solubilities and Solutions

## CHAPTER 12: Nature and Properties of Solution

**Exercise 12.1 (p.207)**

Complete the **dissociation equations** for the following ionic compounds in water.

1. $ZnCl_{2(s)} \xrightarrow{\text{dissolves in water}} Zn^{2+}_{(aq)} + 2\,Cl^-_{(aq)}$

2. $AlCl_{3(s)} \xrightarrow{\text{dissolves in water}} Al^{3+}_{(aq)} + 3\,Cl^-_{(aq)}$

3. $Al_2(SO_4)_{3(s)} \xrightarrow{\text{dissolves in water}} 2\,Al^{3+}_{(aq)} + 3\,SO_4^{2-}_{(aq)}$

4. $Ca(HCO_3)_{2(s)} \xrightarrow{\text{dissolves in water}} Ca^{2+}_{(aq)} + 2\,HCO_3^-_{(aq)}$

5. $Mg(BrO_3)_{2(s)} \xrightarrow{\text{dissolves in water}} Mg^{2+}_{(aq)} + 2\,BrO_3^-_{(aq)}$

6. $K_2CO_{3(s)} \xrightarrow{\text{dissolves in water}} 2K^+_{(aq)} + CO_3^{2-}_{(aq)}$

7. $FeBr_{3(s)} \xrightarrow{\text{dissolves in water}} Fe^{3+}_{(aq)} + 3\,Br^-_{(aq)}$

**Exercise 12.2 (p.207)**

Classify the following mixtures as either **homogeneous** or **heterogeneous.**

(a) a bronze spear

(b) alcoholic beverage

(c) smog

(d) tomato juice

(e) a 12K gold ring

(f) distilled water

(g) muddy water

(h) salad dressing

(i) coca cola

(j) pizza

(k) jello

(l) air

**Solutions**

(a) homogeneous  (b) homogeneous  (c) heterogeneous  (d) heterogeneous  (e) homogeneous

(f) homogeneous  (g) heterogeneous  (h) heterogeneous  (i) homogeneous  (j) heterogeneous

(k) homogeneous  (l) heterogeneous

## Calculations involving concentration of solutions

When solving problems involving concentration of solutions, the following triangle is referred to, as it facilitates the derivations of the relevant equations that are used in the solution process.

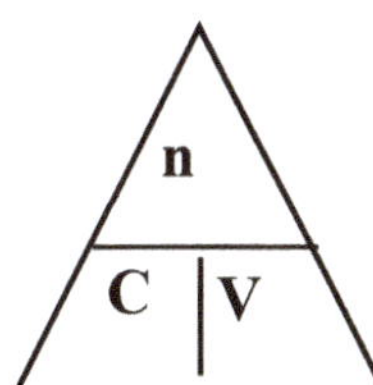


| In this triangle: |
| --- |
| **n = number of moles** |
| **C = concentration** |
| **V = volume** |

The following are the three equations that are derived and used.

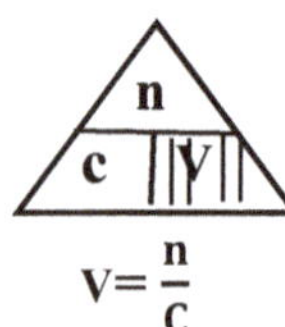

$$V = \frac{n}{C}$$

$$n = C \times V$$

$$C = \frac{n}{V}$$



## Exercise 13.1 (p.211)

Find the molar concentrations of the following solutions:

(a) 0.50 mol of HCl in 500.0 mL of solution

(b) 0.250 mol of $H_2SO_4$ in 200.0 mL of solution

(c) 2.00 mol of NaCl in 1.50 L of solution

(d) 0.10 mol of NaOH in 250.0 mL of solution

(e) 0.20 mol of Ca $(NO_3)_2$ in 1.0L of solution

**Soluitions**

In each of the problems above, the number of moles of solutes and the volumes of the solutions are given. The triangle that is used to derive the appropriate equation for calculating concentration in these problems is as shown on the right.

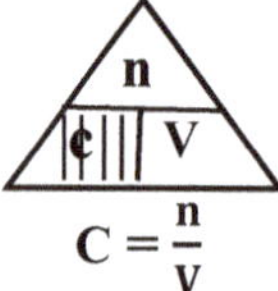

$$C = \frac{n}{V}$$

*Problem:*

**(a)** Find the concentration of a solution that has 0.50 mol of HCl dissolved in 500.0 mL of solution.

*Provided:*

Number of moles (n) of $HCl_{(aq)}$ = 0.50 mol

Volume of solution (V) = 500.0 mL $= \dfrac{500.0 \text{ mL}}{1000 \text{ L/mL}} = 0.5000$ L

**Solution:**

Using the formula, $\mathbf{C = \dfrac{n}{V}}$ we get $\mathbf{C} = \dfrac{0.50 \text{ mol}}{0.5000 \text{ L}} = \mathbf{1.0 \text{ mol/L } HCl_{(aq)}}$

*Problem:*

**(b)** Find the concentration of a solution that has 0.250 mol $H_2SO_{4(aq)}$ dissolved in 200.0 mL of   solution.

*Provided:*

Number of moles (**n**) of $H_2SO_{4(aq)}$ = 0.250 mol

Volume of solution (V) = 200.0 mL $= \dfrac{200.0 \text{ mL}}{1000 \text{ L/mL}} = 0.2000$ L

**Solution:**

Using the formula, $\mathbf{C = \dfrac{n}{V}}$ we get $\mathbf{C} = \dfrac{0.250 \text{ mol}}{0.2000 \text{ L}} = \mathbf{1.25 \text{ mol/L } H_2SO_{4(aq)}}$

***Problem:***

**(c)** Find the concentration of a solution that has 2.00 mol $NaCl_{(aq)}$ dissolved in 1.50 L of solution.

*Provided:*

Number of moles (**n**) of $NaCl_{(aq)}$ = 2.00 mol

Volume of solution (**V**) = 1.50 L

**Solution:**

Using the formula, $C = \dfrac{n}{V}$ we get $C = \dfrac{2.00 \text{ mol}}{1.50 \text{ L}} = $ **1.33 mol/L $NaCl_{(aq)}$**

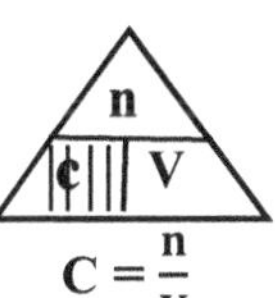

***Problem:***

**(d)** Find the concentration of a solution that has 0.10 mol $NaOH_{(aq)}$ of dissolved in 250.0 mL of solution.

*Provided:*

Number of moles (**n**) of $NaOH_{(aq)}$ = 0.10 mol

Volume of solution (**V**) = 250.0 mL = $\dfrac{250.0 \text{ mL}}{1000 \text{ L/mL}} = 0.2500$ L

**Solution:**

Using the formula, $C = \dfrac{n}{V}$ we get $C = \dfrac{0.10 \text{ mol}}{0.2500 \text{ L}} = $ **0.40 mol/L $NaOH_{(aq)}$**

***Problem:***

**(e)** Find the concentration of a solution that has 0.20 mol $Ca(NO_3)_{2(aq)}$ dissolved in 1.0 L of solution.

*Given:*

Number of moles (**n**) of $Ca(NO_3)_{2(aq)}$ = 0.50 mol

Volume of solution (**V**) = 1.0 L

**Solution:**

Using the formula, $C = \dfrac{n}{V}$ we get $C = \dfrac{0.20 \text{ mol}}{1.0 \text{ L}} = $ **0.20 mol/L $Ca(NO_3)_{2(aq)}$**

***Problem:***

**(f)** Find the concentration of a solution that has 0.0010 mol $KMnO_{4(aq)}$ dissolved in 100.0 mL of solution.

*Given:*

Number of moles (**n**) of $KMnO_{4(aq)}$ = 0.0010 mol

Volume of solution (**V**) = 100.0 mL = $\dfrac{100.0 \text{ mL}}{1000 \text{ L/mL}} = 0.1000$ L

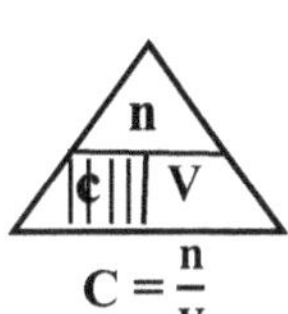

**Solution:**

Using the formula, $C = \dfrac{n}{V}$ we get $C = \dfrac{0.0010 \text{ mol}}{0.1000 \text{ L}} = $ **0.0100 mol/L $KMnO_{4(aq)}$**

When the following problems, there is an initial additional step that must be carried out, when compared to solving those in the previous exercise.  In the previous exercise, the number of moles and voulmes of the compounds were given, where as, in the following ones, the masses and volumes are given. To solve these problems, it is first necessary to convert amounts in mass to amounts in moles.

### *Problem:*

(a) Find the molar concentration of a solution that has 10.0 g of NaOH dissolved in 500.0 mL of solution.

*Provided:*

Mass of $NaOH_{(s)}$ = 10.0 g

Volume of solution (**V**) = 500.0 mL = $\dfrac{500.0 \text{ mL}}{1000 \text{ L/mL}}$ = 0.5000 L

*Required:*

The concentration of the solution

*Steps to be taken:*

**Step 1.**  Find the molar mass of NaOH.

M = 1 × 22.99 g/mol Na + 1 × 16.00 g/mol O + 1 × 1.01 g/mol H = *40.0 g/mol*

**Step 2.**  Find the number of moles of $NaOH_{(aq)}$.

Using the formula, $\mathbf{n} = \dfrac{\mathbf{m}}{\mathbf{M}}$ we get $\mathbf{n} = \dfrac{10.0 \text{ g}}{40.0 \text{ g/mol}}$ = **0.250 mol** $NaOH_{(aq)}$

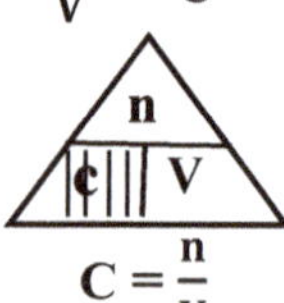


**Step 3.** Find the concentration of the solution. Using the formula, $\mathbf{C} = \dfrac{\mathbf{n}}{\mathbf{V}}$ we get

$$C = \dfrac{0.250 \text{ mol}}{0.5000 \text{ L}} = \textbf{0.500 mol/L } NaOH_{(aq)}$$



### *Problem:*

(b) Find the molar concentration of a solution that has 12.01 g of $H_2SO_{4(aq)}$ dissolved in 800.0 mL of solution.

*Provided:*

Mass of $H_2SO_4$ = 12.01 g

Volume of solution (V) = 800.0 mL = $\dfrac{800.0 \text{ mL}}{1000.0 \text{ L/mL}}$ = 0.800 L

*Required:*

The concentration of the solution.

*Steps to be taken:*

**Step 1.**  Find the molar mass of $H_2SO_{4(aq)}$.

M = 2 × 1.01 g/mol H + 1 × 32.07g/mol S + 4 × 16.00 g/mol S = *98.09 g/mol*

**Step 2**. Find the number of moles of $H_2SO_{4(aq)}$

Using the formula, $n = \dfrac{m}{M}$ we get

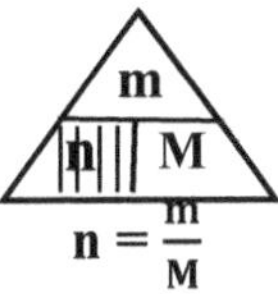

$$n = \dfrac{12.01 \text{ g}}{98.09 \text{ g/mol}} = 0.1224 \text{ mol } H_2SO_{4(aq)}$$

**Step 3.** Find the concentration of the solution. Using the formula, $C = \dfrac{n}{V}$ we get

$$C = \dfrac{0.1224 \text{ mol}}{0.8000 \text{ L}} = 0.153 \text{ mol/L } H_2SO_{4(aq)}$$

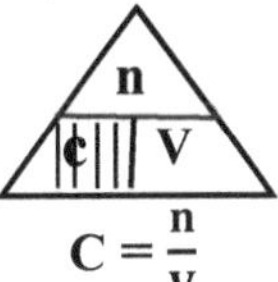

## *Problem:*

(c) Find the molar concentration of a solution that has 6.30 g of $HNO_{3(aq)}$ dissolved in 50.0 mL of solution.

*Provided:*

Mass of $HNO_3$ = 6.30 g

$$\text{Volume of solution (V)} = 50.0 \text{ mL} = \dfrac{50.0 \text{ mL}}{1000 \text{ L/mL}} = 0.0500 \text{ L}$$

*Required:*

The concentration of the solution.

*Steps to be taken:*

**Step 1.** Find the molar mass of $HNO_{3(aq)}$.

$M = 1 \times 1.01 \text{ g/mol H} + 1 \times 14.01 \text{ g/mol N} + 3 \times 16.00 \text{ g/mol} = \textbf{\textit{63.02 g/mol}}$

**Step 2.** Find the number of moles of $HNO_{3(aq)}$.

Using the formula, $n = \dfrac{m}{M}$ we get $n = \dfrac{6.30 \text{ g}}{63.02 \text{g/mol}} = \textbf{0.100 mol } HNO_{3(aq)}$

**Step 3.** Find the concentration of the solution. Using the formula, $C = \dfrac{n}{V}$ we get

$$C = \dfrac{0.100 \text{ mol}}{0.050 \text{ L}} = \textbf{2.00 mol/L } HNO_{3(aq)}$$

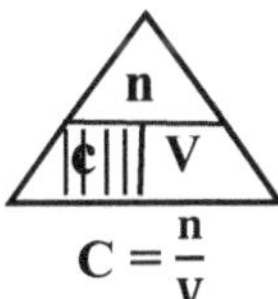

## *Problem:*

(d) Find the molar concentration of a solution that has 9.10 g of $HCl_{(g)}$ dissolved in 200.0 mL of solution.

*Provided:*

Mass: 9.10 g of $HCl_{(g)}$

$$\text{Volume: } 200.0 \text{ mL} = \dfrac{200.0 \text{ mL}}{1000.0 \text{ mL/L}} = 0.2000 \text{ L}$$

*Required:*

The concentration of the solution

*Steps to be taken:*

**Step 1.** Find the molar mass of HCl.

. $M = 1 \times 1.01$ g/mol H $+ 1 \times 35.45$ g/mol Cl $= \mathbf{\textit{36.46 g/mol}}$

**Step 2.** Find the number of moles of HCl.

Using the formula, $\mathbf{n = \dfrac{m}{M}}$ we get $n = \dfrac{9.10 \text{ g}}{36.46 \text{ g/mol}} = \mathbf{0.250}$ **mol** $\text{HCl}_{\text{(aq}}$

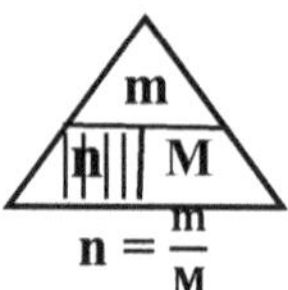

**Step 3.** Find the concentration of the solution. Using the formula, $\mathbf{C = \dfrac{n}{V}}$ we get

$$C = \frac{0.250 \text{ mol}}{0.200 \text{ L}} = \mathbf{1.25 \text{ mol/L } HCl_{(aq)}}$$

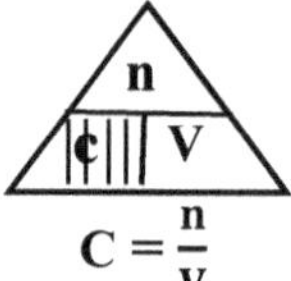

## Problem:

**(e)** Find the molar concentration of a solution that has 2.61 g of $Ba(NO_3)_2$ dissolved in 2.0 L of solution.

*Provided:*

Mass of $Ba(NO_3)_2 = 2.61$ g

Volume $(\mathbf{V}) = 2.0$ L

*Required:*

The concentration of the solution

*Steps to be taken:*

**Step 1.** Find the molar mass of $Ba(NO_3)_2$.

$M = 1 \times 137.33$ g/mol Ba $+ 2 \times 14.01$ g/mol N $+ 6 \times 16.00$ g/mol O $= \mathbf{\textit{261.35 g/mol}}$

**Step 2.** Find the number of moles of $Ba(NO_3)_2$.

Using the formula, $\mathbf{n = \dfrac{m}{M}}$ we get $n = \dfrac{2.61 \text{ g}}{261.35 \text{g/mol}} = \mathbf{0.0100}$ **mol Ba** $\mathbf{(NO_3)_2}$

**Step 3.** Find the concentration of the solution. Using the formula, $\mathbf{C = \dfrac{n}{V}}$ we get

$$C = \frac{0.0100 \text{ mol}}{2.0 \text{ L}} = \mathbf{5.0 \times 10^{-3} \text{ mol/L Ba } (NO_3)_{2(aq)}}$$

## Problem:

**(f)** Find the molar concentration of a solution that has 1.70 g of $AgNO_3$ dissolved in 500.0 mL of solution.

*Provided:*

Mass of $AgNO_3 = 1.70$ g

Volume $(\mathbf{V}) = 500.0$ mL $= \dfrac{500.0 \text{ mL}}{1000 \text{ mL/L}} = 0.5000$ L

*Required:*

The concentration of the solution

*Steps to be taken:*

**Step 1.** Find the molar mass of $AgNO_3$.

$M = 1 \times 107.87 \text{ g/mol Ag} + 1 \times 14.01 \text{ g/mol N} + 3 \times 16.00 \text{ g/mol O} = \textbf{\textit{169.9 g/mol}}$

**Step 2.** Find the number of moles of $AgNO_3$.

Using the formula, $\mathbf{n = \dfrac{m}{M}}$ we get $\mathbf{n} = \dfrac{1.70 \text{ g}}{169.9 \text{ g/mol}} = \textbf{0.0100 mol } AgNO_3.$

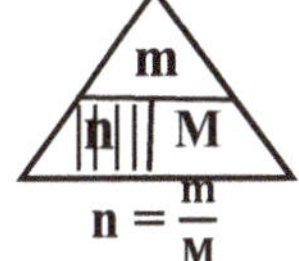

**Step 3.** Find the concentration of the solution. Using the formula, $\mathbf{C = \dfrac{n}{V}}$ we get

$$C = \dfrac{0.0100 \text{ mol}}{0.500 \text{ L}} = \textbf{0.0200 mol/L } AgNO_{3(aq)}$$

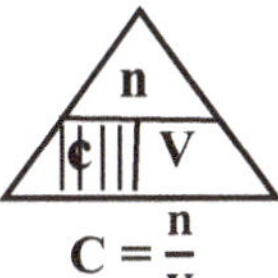



**Finding the masses of solutes that are required to prepare each of the following solutions:**

In the following problems, the volumes and molar concentrations of the soutions are given. The following triangle gives the equation that is used in calculating the number of moles.

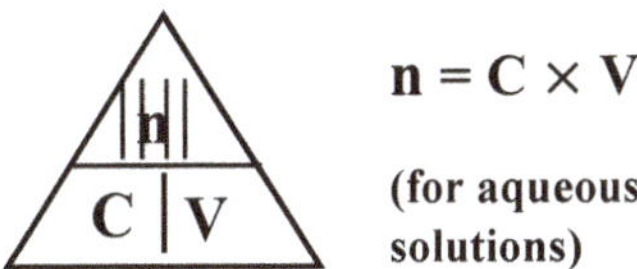

$\mathbf{n = C \times V}$

(for aqueous solutions)

*Problem:*

**(a)** Find the mass of solute that is required to prepare 400.0 mL of 0.200 mol/L $H_2SO_{4(aq)}$.

*Provided:*

Concentration of $H_2SO_{4(aq)}$ = 0.200 mol/L

Volume (**V**) of solution = 400.0 mL = $\dfrac{400.0 \text{ mL}}{1000 \text{mL/L}} = 0.4000 \text{ L}$

*Required:*

The mass of $H_2SO_{4(aq)}$

*Steps to be taken:*

**Step 1.** Find the number of moles of $H_2SO_{4(aq)}$.

Using the formula, $\mathbf{n = C \times V}$ we get

$\mathbf{n} = 0.200 \text{ mol/L} \times 0.4000 \text{ L} = \textbf{0.0800 mol } H_2SO_{4(aq)}$

**Step 2.** Convert amount of $H_2SO_4$ from number of moles to amount in mass. To do this, first find its molar mass.

$M = 2 \times 1.01 \text{ g/mol H} + 1 \times 32.07 \text{ g/mol S} + 4 \times 16.00 \text{ g/mol O} = \textbf{\textit{98.09 g/mol}}$

Using the formula, $\mathbf{m = n \times M}$ we get

$\text{Mass} = 0.0800 \text{ mol} \times 98.09 \text{ g/mol} = \textbf{7.85 g } H_2SO_4$

*Problem:*

**(b)** Find the mass of solute that is required to prepare 200.0 mL of 0.500 mol/L $HCl_{(aq)}$.

*Provided:*

Concentration of $HCl_{(aq)}$ = 0.500 mol/L

Volume (V) of solution = 200.0 mL = $\dfrac{200.0 \text{ mL}}{1000 \text{ mL/L}}$ = 0.2000 L

*Required:*

Mass of NaCl

*Steps to be taken:*

**Step 1**. Find the number of moles of NaCl

Using the formula, **n = C × V** we get **n** = 0.100 mol/L × 0.4000 L

$$= \textbf{0.0400 mol NaCl}$$

**Step 2.** Convert amount of f NaCl from number of moles to amount in mass. To do this, first find    its molar mass.

M = 1 × 22.99 g/mol Na + 1 × 35.45 g/mol Cl = *58.44 g/mol*

Using the formula, **m = n × M** we get

$$\text{Mass} = 0.0400 \text{ mol} \times 58.44 \text{ g/mol}$$

$$= \textbf{2.34 g NaCl}$$

*Problem:*

**(c)** Find the mass of solute that is required to prepare 200.0 mL of 0.500 mol/L $HCl_{(aq)}$.

*Provided:*

Concentration: 0.500 mol/L $HCl_{(aq)}$

Volume: 200.0 mL = $\dfrac{200.0 \text{ mL}}{1000 \text{ mL/L}}$ = 0.2000 L

*Required:*

The mass of $HCl_{(aq)}$

*Steps to be taken:*

**Step 1**. Find the number of moles of $HCl_{(aq)}$.

Using the formula, **n = C × V** we get

**n** = 0.500 mol/L × 0.2000 L = **0.100 mol $HCl_{(aq)}$**

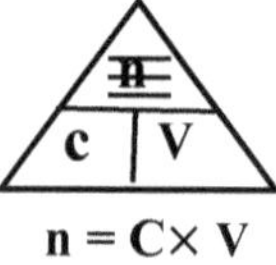


**Step 2.** Convert amount of $HCl_{(aq)}$ from number of moles to amount in mass. To do this, first find the molar mass of HCl.

M = 1 × 1.01 g/mol H + 1 × 35.45  g/mol Cl = *36.46 g/mol*

Using the formula, **m = n × M** we get

Mass = 0.100 mol × 36.46 g/mol = **3.64 g $HCl_{(aq)}$**

*Problem:*

**(d)** Find the mass of solute that is required to prepare 500.0 mL of 0.050 mol/L $KMnO_{4(aq)}$.

*Provided:*

Concentration of $KMnO_{4(aq)}$ = 0.500 mol/L

Volume: $500.0 \text{ mL} = \dfrac{500.0 \text{ mL}}{1000 \text{ mL/L}} = 0.5000 \text{ L}$

*Required:* The mass of $KMnO_4$

*Steps to be taken:*

**Step 1.** Find the number of moles of $KMnO_{4(aq)}$.

Using the formula, **n = C × V** we get

**n** = 0.0500 mol/L × 0.2000 L = **0.100 mol KMnO$_{4(aq)}$**

**Step 2.** Convert amount of $KMnO_{4(aq)}$ from number of moles to amount in mass. To do this, first find its molar mass.

M = 1 × 39.10 g/mol K + 1 × 54.94 g/mol Mn + 4 × 16.00 g/mol O = *158.0 g/mol*

Using the formula, **m = n × M** we get

Mass = 0.100 mol × 158.0 g/mol = **15.8 g KMnO$_{4(aq)}$**

*Problem:*

**(e)** Find the mass of solute that is required to prepare 1.00 L of 0.250 mol/L $Na_2CO_{3(aq)}$.

*Provided:*

Concentration of = 0 .250 mol/L

Volume (**V**) = 1.00 L

*Required:*

The mass of $Na_2CO_3$

*Steps to be taken:*

**Step 1.** Find the number of moles of $Na_2CO_3$.

Using the formula, **n = C × V** we get

**n** = 0.250 mol/L × 1.00 L = **0.250 mol Na$_2$CO$_{3(aq)}$**

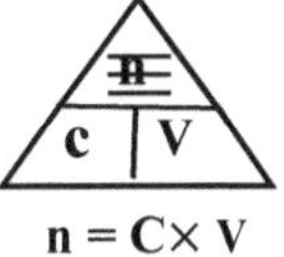


**Step 2.** Convert amount of $Na_2CO_{3(aq)}$ from number of moles to amount in mass. To do this, first find its molar mass.

M = 2 × 22.99 g/mol Na + 1 × 12.01 g/mol C + 3 × 16.00 g/mol O = *106.0 g/mol*

Using the formula, **m = n × M** we get

Mass = 0.250 mol × 106.0 g/mol = **26.5 g Na$_2$CO$_3$**

## Preparing solutions of hydrated salts

Find the mass of solute that is required to prepare each of the following solutions:

Solving the following problems follow the same strategies as those in **Exercise 13.3**, except that the mass of the water of crystallization must be included.

### *Problem:*

(a) What mass of $CaCl_2 \cdot 2H_2O_{(S)}$ is required to prepare 400.0 mL of a 0.200 mol/L $CaCl_{2(aq)}$ solution?

*Provided:*

Concentration of $CaCl_{2(aq)}$ = 0.200 mol/L

$$\text{Volume of } CaCl_{2(aq)} = 400.0 \text{ mL} = \frac{400.0 \text{ mL}}{1000 \text{ mL/L}} = 0.4000 \text{ L}$$

*Required:* The mass of $CaCl_2 \cdot 2H_2O_{(S)}$

*Steps to be taken:*

**Step 1.** Find the number of moles of $CaCl_2 \cdot 2H_2O_{(S)}$

Using the formula, $\mathbf{n = C \times V}$ we get

$\mathbf{n}$ = 0.200 mol/L $\times$ 0.4000 L = **0.0800 mol $CaCl_2 \cdot 2H_2O_{(S)}$**

**Step 2.** Convert amount of $CaCl_2 \cdot 2H_2O_{(S)}$ from number of moles to amount in mass. To do this, first find the molar masses of $CaCl_2$ and $H_2O$ respectively.

$M = 1 \times 40.08$ g/mol Ca $+ 2 \times 35.45$ g/mol Cl = ***110.0 g/mol CaCl₂***

$M = 2 \times 1.01$ *g/mol* $+ 1 \times 16.00$ g/mol = ***18.02 g/mol H₂O***

Using the formula, $\mathbf{m = n \times M}$ we get for the mass of $CaCl_2$

Mass = 0.0800 mol $\times$ 110.0 g/mol = **8.80 g CaCl₂**

*Note that for every mole of $CaCl_2$, there are **2 moles** of $H_2O$. Therefore, in 0.080 moles

of $CaCl_2 \cdot 2H_2O$, the number of moles of $H_2O$ will be $= 2 \times 0.0800$ mol = **0.160 mol H₂O**.

Using the formula, $\mathbf{m = n \times M}$ we get for the mass of $H_2O$

Mass = 0.160 mol $\times$ 18.02 g/mol = **2.88 g H₂O**

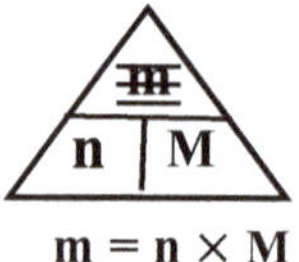

The total mass of $CaCl_2 \cdot 2H_2O_{(S)}$ required to make 400.0 mL of a 0.200 mol/L $CaCl_{2(aq)}$ solution

$=$ **8.80 g** (from $CaCl_2$) **+ 2.88 g** (from $H_2O$) = **11.6 g of $CaCl_2 \cdot 2H_2O$**

### *Problem:*

(b) What mass of $Na_2SO_4 \cdot 10H_2O_{(S)}$ is required to prepare 800.0 mL of a 0.100 mol/L $Na_2SO_{4(aq)}$ solution?

*Provided:*

Concentration $Na_2SO_{4(aq)}$ = 0.100 mol/L

$$\text{Volume of } Na_2SO_{4(aq)} = 800.0 \text{ mL} = \frac{800.0 \text{ mL}}{1000 \text{ mL/L}} = 0.8000 \text{ L}$$

*Required:*

The mass of $Na_2SO_4 \cdot 10H_2O_{(S)}$

*Steps to be taken:*

**Step 1.** Find the number of moles of $Na_2SO_4 \cdot 10H_2O$

Using the formula, **n = C × V** we get

$$n = 0.100 \text{ mol/L} \times 0.8000 \text{ L}$$

$$= \textbf{0.0800 mol } Na_2SO_4 \cdot 10H_2O_{(aq)}$$

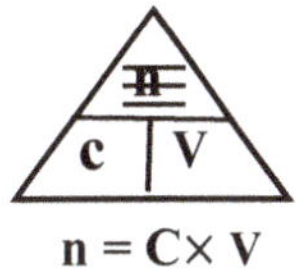


**Step 2.** Convert amount of $Na_2SO_4 \cdot 10H_2O$ from number of moles to amount in mass. To do this step, *the molar masses of $Na_2SO_4$ and $H_2O$ must be first calculated.*

$$M_{Na_2SO_4} = 2 \times 22.99 \text{ g/mol Na} + 1 \times 32.07 \text{ g/mol S} + 4 \times 16.00 \text{ g/mol Cl} = \textbf{\textit{142.1 g/mol } Na_2SO_4}$$

$$M_{H_2O} = 2 \times 1.01 + 1 \times 16.00 \text{ g/mol} = \textbf{\textit{18.02 g/mol } H_2O}$$

Using the formula, **m = n × M** we get for the mass of **$Na_2SO_4$**

$$\text{Mass} = 0.0800 \text{ mol} \times 142.1 \text{ g/mol} = \textbf{11.4 g } Na_2SO_4$$

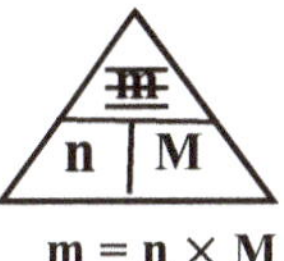


- Note that for every mole of $CaCl_2$, there are ten moles of $H_2O$. Therefore, in 0.080 moles of $Na_2SO_4 \cdot 10H_2O$, the number of moles of $H_2O$ will be

$$= 10 \times 0.0800 \text{ mol} = \textbf{0.800 mol } H_2O.$$

Using the formula, **m = n × M** we get for the mass of $H_2O$

$$\text{Mass} = 0.800 \text{ mol} \times 18.02 \text{ g/mol} = \textbf{14.4 g } H_2O$$



- The total mass of $Na_2SO_4 \cdot 10H_2O$ required to make 800.0 mL of a 0.100 mol/L solution

$$= \textbf{11.4 g} \text{ (from } Na_2SO_4\text{)} + \textbf{14.4 g} \text{ (from } H_2O\text{)} = \textbf{25.8 g } Na_2SO_4 \cdot 10H_2O$$

| **Exercise 13.5 (p.217)** | **Exercise 13.6 (p.217)** |
|---|---|

**a)** $NaOH_{(aq)} \rightarrow Na^+_{(aq)} + OH^-_{(aq)}$

   1 mol/L     1 mol/L    1 mol/L

**b)** $ZnSO_{4(aq)} \rightarrow Zn^{2+}_{(aq)} + SO_4^{2-}_{(aq)}$

   1 mol/L     1 mol/L    1 mol/L

**c)** $CaCl_{2(aq)} \rightarrow Ca^{2+}_{(aq)} + 2\,Cl^-_{(aq)}$

   1 mol/L     1 mol/L    2 mol/L

**d)** $Mg(OH)_{2(aq)} \rightarrow Mg^{2+}_{(aq)} + 2(OH^-)_{(aq)}$

   1 mol/L     1 mol/L    2 mol/L

**f)** $FeCl_{3(aq)} \rightarrow Fe^{3+}_{(aq)} + 3\,Cl^-_{(aq)}$

   1 mol/L     1 mol/L    3 mol/L

**g)** $Al_2(SO_4)_{3(aq)} \rightarrow 2\,Al^{3+}_{(aq)} + 3SO_4^{2-}_{(aq)}$

   1 mol/L     2 mol/L    3 mol/L

**a)** $HCl_{(aq)} \rightarrow H^+_{(aq)} + Cl^-_{(aq)}$

   1 mol/L     1 mol/L    1 mol/L

**b)** $HNO_{3(aq)} \rightarrow H^+_{(aq)} + NO_3^-_{(aq)}$

   1 mol/L     1 mol/L    1 mol/L

**c)** $HI_{(aq)} \rightarrow H^+_{(aq)} + I^-_{(aq)}$

   1 mol/L     1 mol/L    1 mol/L

**d)** $HCN_{(aq)} \rightarrow H^+_{(aq)} + CN^-_{(aq)}$

   1 mol/L     1 mol/L    1 mol/L

**e)** $HClO_{3(aq)} \rightarrow H^+_{(aq)} + ClO_3^-_{(aq)}$

   1 mol/L     1 mol/L    1 mol/L

## Concentration of ions

Assuming complete dissociation, calculate the molar concentration of all ions in the following solutions that have:

***Problem*:**

   **(a)** Calculate the molar concentration of all ions in a solution that has 2.49 g of $CuSO_4 \bullet 5H_2O$ $_{(s)}$ dissolved in 400.0 mL of solution.

*Provided:*

    Mass of $= CuSO_4 \bullet 5H_2O_{(s)} = 2.49$ g

    Volume of solution (V) $= 400.0$ mL $= \dfrac{400.0 \text{ mL}}{1000 \text{ mL/L}} = 0.4000$ L

*Required:*

    The molar concentrations of $Cu^{2+}_{(aq)}$ and $SO_4^{2-}_{(aq)}$

*Steps to be taken:*

   **Step 1.** Convert amount of $CuSO_4 \bullet 5H_2O$ from number of moles to amount in mass to amount in number of moles. To do this, first calculate the molar masses of $CuSO_4$ and $H_2O$.

   M $_{CuSO_4}$ = 1 × 63.55 g/mol Cu + 1 × 32.07 g/mol S + 4 × 16.00 g/mol O = ***159.6 g/mol***

    •   Note that for every mole of $CuSO_4 \bullet 5H_2O$ there are 5 moles of $H_2O$.

   5 M$_{H_2O}$ = 5 × 2.02 g/mol $H_2$ + 5 16.00 g/mol O = ***90.1g/mol $H_2O$***

      Molar mass for $CuSO_4 \bullet 5H_2O$

    = ***159.6 g*** (from 1 mole $CuSO_4$) + ***90.1g*** (from 5 moles $H_2O$) = **249.7 g/mol**

      Using the formula, $\mathbf{n = \dfrac{m}{M}}$ we get

      $\mathbf{n = \dfrac{2.49 \text{ g}}{249.7 \text{ g/mol}} = 0.0100 \text{ mol } CuSO_4 \bullet 5H_2O}$

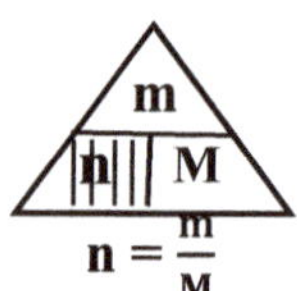

   **Step 2.** Find the concentration of the solution. Using the formula, $\mathbf{C = \dfrac{n}{V}}$ we get

      $\mathbf{C = \dfrac{0.0100 \text{ mol}}{0.4000 L} = 0.0250 \text{ mol/L } CuSO_4 \bullet 5H_2O}$

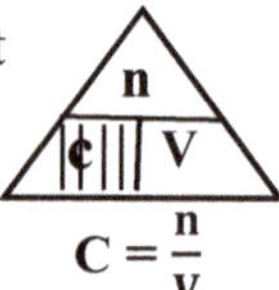

   **Step 3.** Write a balanced equation to show how the compound dissociate in water to give its specific mole ratios of ion concentrations. **Note that only the $CuSO_4$ ionizes;** *water barely ionizes, so its ions are not included here.*

      $CuSO_{4(aq)}$       $\rightarrow$    $Cu^{2+}_{(aq)}$    +    $SO_4^{2-}$

      1 mol/L           1 mol/L         1 mol/L

   **Step 4.** Calculate the molar concentration of each ion.

      $[Cu^{2+}_{(aq)}]$   = $0.0250$ mol/L $CuSO_{4(aq)}$ × $\dfrac{1 \text{ mol/L } Cu^{2+}(aq)}{1 \text{ mol/} CuSO_{4(aq)}}$

            = **0.0250 mol/L**

      $[SO_4^{2-}]$   = $0.0250$ mol/L $CuSO_{4(aq)}$ × $\dfrac{1 \text{ mol/L } SO_4^{2-}}{1 \text{ mol/} CuSO_{4(aq)}}$

            = **0.0250 mol/L**

*Problem:*

**(b)** Calculate the molar concentration of all ions in a solution that has 1.74 g of $Na_2SO_{4(s)}$ dissolved in 500.0 mL of solution.

*Provided:*

Mass of $Na_2SO_{4(s)}$ = 1.74 g

Volume of solution (V) = 500.0 mL = $\dfrac{500.0\ \text{mL}}{1000\ \text{mL/L}}$ = 0.5000 L

*Required:*

The molar concentrations of $Na^+_{(aq)}$ and $SO_4^{2-}_{(aq)}$

*Steps to be taken:*

**Step 1.** Convert amount of $Na_2SO_{4(s)}$ from amount in mass to amount in number of moles. To do this, first calculate its molar mass.

M = 2 × 22.99 g/mol N + 1 × 32.07 g/mol S + 4 × 16.00 g/mol O = *142.1 g/mol*

Using the formula, $\mathbf{n = \dfrac{m}{M}}$ we get

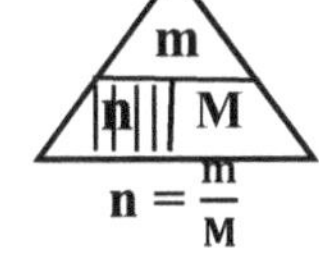

$$n = \dfrac{1.74\ \text{g}}{142.1\ \text{g/mol}} = \textbf{0.0119 mol } Na_2SO_{4(s)}$$

**Step 2.** Find the concentration of the solution.

Using the formula, $\mathbf{C = \dfrac{n}{V}}$ we get

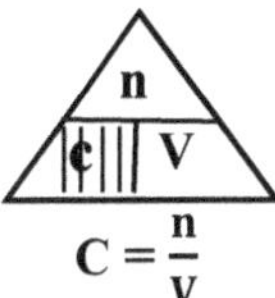

$$C = \dfrac{0.0119\ \text{mol}}{0.5000\ \text{L}} = \textbf{0.0238 mol/L } Na_2SO_{4(aq)}$$

**Step 3.** Write a balanced equation to show how the compound, $\mathbf{Na_2SO_{4(aq)}}$ dissociates in water to give its specific mole ratios of ion concentrations.

$$Na_2SO_{4(aq)} \quad\longrightarrow\quad 2\ Na^+_{(aq)} \quad + \quad SO_4^{2-}$$

| | | |
|---|---|---|
| 1 mol/L | 2 mol/L | 1 mol/L |

**Step 4.** Calculate the molar concentration of each ion.

$$[Na^{2+}_{(aq)}] = 0.0238\ \text{mol/L } Na_2SO_{4(aq)} \times \dfrac{2\ \text{mol/L } Na^+(aq)}{1\ \text{mol/}Na_2SO_{4(aq)}}$$

$$= \textbf{0.0476 mol/L}$$

$$[SO_4^{2-}] = 0.0238\ \text{mol/L } Na_2SO_{4(aq)} \times \dfrac{1\ \text{mol/L } SO_4^{2-}_{(aq)}}{1\ \text{mol/}Na_2SO_{4(aq)}}$$

$$= \textbf{0.0238 mol/L}$$

*Problem:*

**(c)** Calculate the molar concentration of all ions in a solution that has 11.10 g of $CaCl_{2(s)}$ dissolved in 500.0 mL of solution.

*Given:*

Mass of $CaCl_{2(s)}$ = 1.74 g

Volume of solution (V) = 500.0 mL = $\dfrac{500.0\ \text{mL}}{1000\ \text{mL/L}}$ = 0.5000 L

*Required:*

The molar concentrations of $Ca^{2+}_{(aq)}$ and $Cl^-_{(aq)}$

*Steps to be taken:*

**Step 1**. Convert amount of $CaCl_{2(s)}$ from amount in mass to amount in number of moles. To do this, first calculate its molar mass.

$M = 1 \times 40.08$ g/mol Ca $+ 2 \times 35.45$ g/mol Cl $= \mathbf{\textit{111.0 g/mol}}$

Using the formula, $\mathbf{n} = \dfrac{\mathbf{m}}{\mathbf{M}}$ we get

$$\mathbf{n} = \frac{11.10 \text{ g}}{111.0 \text{ g/mol}} = \mathbf{0.1000 \text{ mol } CaCl_{2(s)}}$$

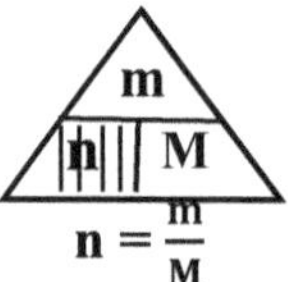


**Step 2.** Find the concentration of the solution.

Using the formula, $\mathbf{C} = \dfrac{\mathbf{n}}{\mathbf{V}}$ we get

$$\mathbf{C} = \frac{0.1000 \text{ mol}}{0.5000 \text{L}} = \mathbf{0.2000 \text{ mol/L } CaCl_{2(aq)}}$$

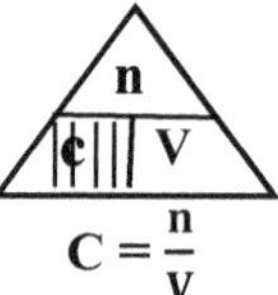


**Step 3.** Write a balanced equation to show how $CaCl_{2(aq)}$ dissociates in water to give its specific mole ratios of ion concentrations.

| $CaCl_{2(aq)}$ | $\rightarrow$ | $Ca^{2+}_{(aq)}$ | $+$ | $2\ Cl^-$ |
|---|---|---|---|---|
| 1 mol/L | | mol/L | | 2 mol/L |

Step 4. Calculate the molar concentration of each ion

$$[Ca^{2+}_{(aq)}] = 0.2000 \text{ mol/L } CaCl_{2(aq)} \times \frac{1 \text{ mol/L } Ca^{2+}(aq)}{1 \text{ mol/}CaCl_2}$$

$$= \mathbf{0.2000 \text{ mol/L}}$$

$$[Cl^-] = 0.2000 \text{ mol/L } CaCl_{2(aq)} \times \frac{2 \text{ mol/L } Cl^-}{1 \text{ mol/}CaCl_2}$$

$$= \mathbf{0.400 \text{ mol/L}}$$

**Problem:**

**d)** Calculate the molar concentration of all ions in a solution that has 1.71 g of $Ba(OH)_{2(s)}$ dissolved in 200.0 mL of solution.

*Provided:*

Mass of $Ba(OH)_{2(s)} = 1.74$ g

Volume of solution $(V) = 200.0$ mL $= \dfrac{200.0 \text{ mL}}{1000 \text{ mL/L}} = 0.2000$ L

*Required:* The molar concentrations of $Ba^{2+}_{(aq)}$ and $OH^-_{(aq)}$

*Steps to be taken:*

**Step 1.** Convert amount of $Ba(OH)_{2(s)}$ from amount in mass to amount in number of moles. To do this, first calculate the molar mass of $Ba(OH)_{2(s)}$.

$$M = 1 \times 137.33 \text{ g/mol Ba} + 2 \times 1.01 \text{ g/mol H} + 2 \times 16.00 \text{ g/mol O} = \mathbf{171.35\ g/mol}$$

Using the formula, $\mathbf{n = \dfrac{m}{M}}$ we get

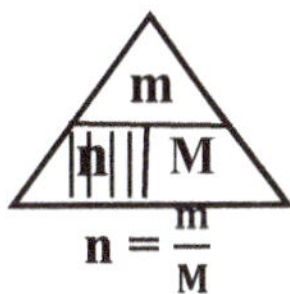

$$\mathbf{n} = \frac{1.71 \text{ g}}{171.35 \text{ g/mol}} = \mathbf{0.0100\ mol\ Ba(OH)_{2(s)}}$$

**Step 2.** Find the concentration of the solution.

Using the formula, $\mathbf{C = \dfrac{n}{V}}$ we get

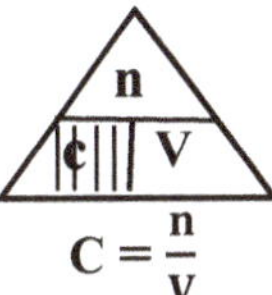

$$C = \frac{0.0100 \text{ mol}}{0.2000 \text{L}} = \mathbf{0.0500\ mol/L\ Ba(OH)_{2(aq)}}$$

**Step 3.** Write a balanced equation to show how the compound, $\mathbf{Ba(OH)_{2(aq)}}$ dissociates in water to give its specific mole ratios of ion concentrations.

$$Ba(OH)_{2(aq)} \quad \longrightarrow \quad Ba^{2+}{}_{(aq)} \quad + \quad 2\ OH^-$$

| | | |
|---|---|---|
| 1 mol/L | 1 mol/L | 2 mol/L |

**Step 4.** Calculate the molar concentration of each ion.

$$[Ba^{2+}{}_{(aq)}] = 0.0500 \text{ mol/L } Ba(OH)_{2(aq)} \times \frac{1 \text{ mol/L } Ba^{2+}aq)}{1 \text{ mol/Ba(OH)2}}$$

$$= \mathbf{0.0500\ mol/L}$$

$$[OH^-] = 0.0500 \text{ mol/L } Ba(OH)_{2(aq)} \times \frac{2 \text{ mol/L } OH^-}{1 \text{ mol/Ba(OH)}_2}$$

$$= \mathbf{0.100\ mol/L}$$

**Exercise 13.8 (p.222)**

## Dilution of Solutions

### *Problem:*

(a) What is the final concentration of a 200.0 mL of 0.100 mol/L $HCl_{(aq)}$, if it is diluted to a new volume of 800.0 mL?

### *Provided:*

Initial concentration, $C_1 = 0.100$ mol/L $HCl_{(aq)}$

$$\text{Initial volume, } V_1 = 200 \text{ mL} = \frac{200.0 \text{ mL}}{1000 \text{ mL/L}} = 0.2000 \text{ L}$$

$$\text{Final volume, } V_2 = 800.0 \text{ mL} = \frac{800.0 \text{ mL}}{1000 \text{ mL/L}} = 0.8000 \text{ L}$$

### *Required:*

Final concentration, $C_2$

*Steps to be taken:*

**Step 1.** Using the ion $C_1 \times V_1 = C_2 \times V_2$, make $C_2$ the subject of the equation as follows:

$$C_2 = \frac{C_1 \times V_1}{V_2} \quad \text{Substituting the values in this equation we get}$$

$$C_2 = \frac{0.100\frac{mol}{L} \times 0.2000L}{0.8000\ L} = \textbf{0.0250 mol/L } HCl_{(aq)}$$

***Problem:***

(**b**) A 100.0 mL of a solution of $NaOH_{(aq)}$ of was diluted to a new volume of 1.0 L. If its final concentration is 0.01 mol/L, determine its initial concentration.

*Provided:*

Final concentration, $C_2$ = 0.01 mol/L $NaOH_{(aq)}$

Initial volume, $V_1$ = 100.0 mL = $\dfrac{100.0\ mL}{1000\ mL/L}$ = 0.1000 L

Final volume, $V_2$ = 1.0 L

*Required:*

Initial concentration, $C_1$

*Steps to to be taken:*

**Step 1.** Using the ion $C_1 \times V_1 = C_2 \times V_2$, make $C_1$ the subject of the equation as follows:

$$C_1 = \frac{C_2 \times V_2}{V_1} \quad \text{Substituting the values in this equation we get}$$

$$C_1 = \frac{0.01\frac{mol}{L} \times 1.0L}{0.1000/L} = \textbf{0.10 mol/L}$$

***Problem:***

(**c**) A student needs to make 2.0 L of 0.10 mol/L of $H_2SO_{4(aq)}$ by diluting a sample of a stock solution of 10.0 mol/L C. Determine the volume of the stock solution must be diluted. What precautions must be taken when diluting concentrated acids?

*Provided:*

Final concentration, $C_2$ = 0.01 mol/L of $H_2SO_{4(aq)}$

Initial concentration, $C_1$ = 10.0 mol/L

Final volume, $V_2$ = 1.0 L

*Required:*

Initial volume: $V_1$

*Steps to be taken:*

**Step 1.** Using the ion $C_1 \times V_1 = C_2 \times V_2$, make $V_1$ the subject of the equation as follows:

$$V_1 = \frac{C_2 \times V_2}{C_1} \quad \text{Substituting the values in this equation we get}$$

$$V_1 = \frac{0.01\frac{mol}{L} \times 1.0L}{10.0\ mol/L} = 0.0010\ L = \textbf{1.00 mL}$$

- When diluting concentrated acids, **never add water to the acids, but always add the acids to water,** using a cold-water bath. All protective equipment; goggles, gloves and aprons must be worn when diluting concentrated acids.

### Problem:

**(d)** To what final volume must 10.0 mL of 10.0 mol/L $HCl_{(aq)}$ be diluted for its concentration to change to 0.50 mol/L?

*Provided:*

Final concentration, $C_2$ = 0.50 mol/L

Initial concentration, $C_1$ = 10.0 mol/L

Initial volume $V_1$: 10.0 mL $= \dfrac{10.0 \text{ mL}}{1000 \text{ mL/L}} = 0.0100$ L

*Required:*

Final volume, $V_2$

*Steps to be taken:*

**Step 1.** Using the ion $C_1 \times V_1 = C_2 \times V_2$, make $V_2$ the subject of the equation as fol lows:

$$V_2 = \frac{C_1 \times V_1}{C_2} \text{ Substituting the values in this equation we get}$$

$$V_2 = \frac{10.0\frac{mol}{L} \times 0.0100 \text{ L}}{0.50 \text{ mol/L}} = \textbf{0.20 L = 200 mL}$$

### Problem:

**(d)** An aqueous solution has 12.26 g $H_2SO_{4(aq)}$ dissolved in 500.0 mL of solution. What is the concentration of the solution? What will be the final concentration of $H^+_{(aq)}$ ions and $SO_4^{2-}{}_{(aq)}$ ions if the solution is diluted to 2.00 L?

*Provided:*

Mass of $H_2SO_4$ = 12.26 g

Initial Volume = 500 mL $= \dfrac{500 \text{ mL}}{1000 \text{ mL/L}} = 0.500$ L

Final Volume = 2.0 L

*Required:*

Final molar concentrations of $H^+_{(aq)}$ and $SO_4^{2-}{}_{(aq)}$

**Step 1.** Convert amount of $H_2SO_4$ from number of moles to amount in mass to amount in number of moles. To do this, first find the molar mass of $H_2SO_4$.

$$M = 2 \times 1.01 \text{ g/mol H} + 1 \times 32.07 \text{ g/mol S} + 4 \times 16.00 \text{ g/mol O} = \textbf{\textit{98.09 g/mol}}$$

Using the formula, $\mathbf{n = \dfrac{m}{M}}$ we get

$$n = \frac{12.26 \text{ g}}{98.09 \text{ g/mol}} = \textbf{0.125 mol } \mathbf{H_2SO_4}$$

**Step 2.** Find the concentration of the solution.

Using the formula, $C = \dfrac{n}{V}$ we get

$$C = \frac{0.125 \text{ mol}}{0.5000 \text{L}} = \textbf{0.250 mol/L } H_2SO_4$$

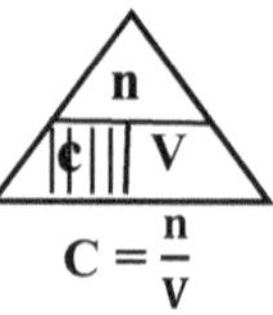

$$C = \frac{n}{V}$$

**Step 3.** Find the final concentration $C_2$ of $H_2SO_{4(aq)}$.

Initial concentration, $C_1 = 0.250$ mol/L

Initial volume, $V_1 = 0.5000$ L

Final volume $V_2 = 2.00$ L

Using the ion $C_1 \times V_1 = C_2 \times V_2$, make $C_2$ the subject of the equation as fol lows:

$$C_2 = \frac{C_1 \times V_1}{V_2}$$ Substituting the values in this equation we get

$$C_2 = \frac{0.250 \frac{\text{mol}}{\text{L}} \times 0.500 \text{ L}}{2.00 \text{ L}} = \textbf{0.0625 mol/L } H_2SO_4$$

**Step 4.** Write a balanced equation to show how $H_2SO_{4(aq)}$ ionizes in water to give its specific mole ratios of ion concentrations. Here we are assuming the $H_2SO_{4(aq)}$ ionizes completely.

$$H_2SO_{4(aq)} \qquad \rightarrow \qquad 2\ H^+_{(aq)} \qquad + \qquad SO_4^{2-}$$

$$1 \text{ mol/L} \qquad\qquad 2 \text{ mol/L} \qquad\qquad 1 \text{ mol/L}$$

**Step 5.** Calculate the molar concentration of each ion.

$$[H^+_{(aq)}] = 0.0625 \text{ mol/L } H_2SO_{4(aq)}. \times \frac{2 \text{ mol/L } H^+(aq)}{1 \text{ mol/}H_2SO_{4(aq)}} = \textbf{0.125 mol/L}$$

$$[SO_4^{2-}] = 0.0625 \text{ mol/L } H_2SO_{4(aq)}. \times \frac{1 \text{ mol/L } SO_4^{2-}_{(aq)}}{1 \text{ mol/}H_2SO_{4(aq)}} = \textbf{0.0625 mol/L}$$

## *Problem:*

**(e)** An aqueous solution has 2.92 g $Mg(OH)_{2(s)}$ dissolved in 200.0 mL of solution. What is the concentration of this solution? What will be the final concentration of $Mg^{2+}_{(aq)}$ ions and $OH^-_{(aq)}$ ions if the solution is diluted to 1.00 L?

## *Provided:*

Mass of = 2.92 g $Mg(OH)_{2(s)}$

Initial Volume: $V_1 = 200.0$ mL $= \dfrac{200.0 \text{ mL}}{1000 \text{ mL/L}} = 0.2000$ L

Final Volume = 1.00 L

## *Required:*

Final molar concentrations of $Mg^{2+}_{(aq)}$ and $OH^-_{(aq)}$

*Steps to be taken:*

**Step 1.** Convert amount of $Mg(OH)_{2(s)}$ from number of moles to amount in mass to amount in number of moles. To do this step, first find its molar mass.

$$M = 1 \times 24.31 \text{ g/mol Mg} + 2 \times 1.01 \text{ g/mol H} + 2 \times 16.00 \text{ g/mol O} = \textbf{\textit{58.33 g/mol}}$$

Using the formula, $\mathbf{n = \dfrac{m}{M}}$ we get

$$\mathbf{n = \dfrac{2.92 \text{ g}}{58.33 \text{ g/mol}} = 0.0500 \text{ mol } Mg(OH)_{2(s)}}$$

**Step 2.** Find the concentration of the solution.

Using the formula, $\mathbf{C = \dfrac{n}{V}}$ we get

$$\mathbf{C = \dfrac{0.0500 \text{ mol}}{0.2000 \text{L}} = 0.250 \text{ mol/L}}$$

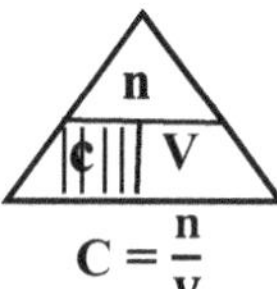

**Step 3.** Find the final concentration, $C_2$ of $Mg(OH)_{2(aq)}$.

Initial concentration, $C_1 = 0.250$ mol/L

Initial volume, $V_1 = 0.2000$ L mol/L

Final volume $V_2 = 1.00$ L

Using the ion $C_1 \times V_1 = C_2 \times V_2$, make $C_1$ the subject of the equation as fol lows:

$$C_2 = \dfrac{C_1 \times V_1}{V_2} \quad \text{Substituting the values in this equation we get}$$

$$C_2 = \dfrac{0.250 \frac{\text{mol}}{\text{L}} \times 0.200 \text{ 1L}}{1.00 \text{ L}} = \textbf{0.0500 mol/L } \mathbf{Mg(OH)_{2(aq)}}.$$

**Step 4.** Write a balanced equation to show how $Mg(OH)_{2(aq)}$ ionizes in water to give its  specific mole ratios of ion concentrations.

$$Mg(OH)_{2(aq)} \quad \rightarrow \quad Mg^{2+}_{(aq)} \quad + \quad (OH^-)$$

| | | |
|---|---|---|
| 1 mol/L | 1 mol/L | 2 mol/L |

**Step 5.** Calculate the molar concentration of each ion in the solution.

$$[Mg^{2+}_{(aq)}] = 0.0500 \text{ mol/L } Mg(OH)_{2(aq)} \times \dfrac{1 \text{ mol/L } Mg^{2+}(aq)}{1 \text{ mol/} Mg(OH)_{2(aq)}} = \textbf{0.0500 mol/}$$

$$[OH^-] = 0.0500 \text{ mol/L } Mg(OH)_{2(aq)} \times \dfrac{2 \text{ mol/L } OH^-}{1 \text{ mol/} Mg(OH)_{2(aq)}} = \textbf{0.100 mol/L}$$

***Problem:***

**(f)**  What mass of $Na_2CO_{3(s)}$ must be dissolved in 200.0 mL of solution to produce a solution having a $CO_3^{2-}{}_{(aq)}$ ions of 0.500 mol/L?

*Given:*

$$\text{Volume: } 200.0 \text{ mL} = \dfrac{200.0 \text{ mL}}{1000 \text{ mL/L}} = 0.2000 \text{ L}$$

Concentration of $CO_3^{2-}{}_{(aq)} = 0.500$ mol/L

*Required:*

The mass of $Na_2CO_{3(s)}$ that must be dissolved in 200.0 mL of solution to produce 0.500 mol/L $CO_3^{2-}{}_{(aq)}$.

*Steps to be taken:*

**Step 1.** Write a balanced equation to show how $Na_2CO_{3(aq)}$ ionizes in water to give its specific mole ratios of ion concentrations.

$$Na_2CO_{3(aq)} \quad\rightarrow\quad 2\ Na^+{}_{(aq)} \quad+\quad CO_3^{2-}{}_{(aq)}$$

$$1\ mol/L \qquad\qquad 2\ mol/L \qquad\qquad 1\ mol/L$$

**Step 2.** Use the concentration of $CO_3^{2-}{}_{(aq)}$ ions to find the concentration of $Na_2CO_{3(aq)}$.

According to the ionization equation in step **1**, 1 mol/ $CO_3^{2-}{}_{(aq)}$ is produced from 1 mol/L $Na_2CO_{3(s)}$. Knowing the concentration of $CO_3^{2-}{}_{(aq)}$, the concentration of $Na_2CO_{3(aq)}$ can thus be calculated according to the following equation:

$$[\,Na_2CO_{3(s)}] \;=\; \frac{1\frac{mol\,Na_2\,CO_3}{L}}{1\ mol/L\,CO_3^{2-}} \;\times\; 0.500\ mol/L\ CO_3^{2-}{}_{(aq)}$$

$$=\; \textbf{0.500 mol/L } \mathbf{Na_2CO_{3(aq)}}$$

**Step 3.** Since the concentration and volume of $Na_2CO_{3(aq)}$ are known, its number of moles can be found using the following equation:

$$\mathbf{n = C \times V} \;=\; 0.500\ mol/L \times 0.2000\ L$$

$$=\; \textbf{0.100 mol } \mathbf{Na_2CO_{3(s)}}$$

$$\boxed{\frac{n}{\;c\;|\;V\;}} \qquad \mathbf{n = C \times V}$$

**Step 4.** Convert amount of $Na_2CO_3$ from number of moles to amount in mass. To do this, first find its molar mass.

$$M = 2 \times 22.99\ g/mol\ Na \quad 1 \times 12.01\ g/mol\ C + 3 \times 16.00\ g/mol\ O \quad = \textit{106 g/mol}$$

Using the formula, $\mathbf{m = n \times M}$ we get

Mass $= 0.100\ mol \times 106\ g/mol = \textbf{10.6 g } \mathbf{Na_2CO_{3(s)}}$

$$\boxed{\frac{m}{\;n\;|\;M\;}} \qquad \mathbf{m = n \times M}$$

## Problem:

**(g)** The concentration of $H^+{}_{(aq)}$ ions is important in determining the pH of solutions. What mass of $HCl_{(aq)}$ must be dissolved in 500.0 mL of solution to produce a concentration of 0.100 mol/L $H^+{}_{(aq)}$ ions?

*Given:*

Volume: $500.0\ mL = \dfrac{500.0\ mL}{1000\ mL/L} = 0.5000\ L$

Concentration of $H^+{}_{(aq)} = 0.100\ mol/L$

*Required:*

The mass of $HCl_{(aq)}$ that must be dissolved in 500.0 mL of solution to produce 0.100 mol/L $H^+{}_{(aq)}$ ions.

*Steps to be taken:*

**Step 1.** Write a balanced equation to show how $HCl_{(aq)}$ ionizes in water to give its specific mole ratios of ion concentrations.

$$HCl_{(aq)} \quad\rightarrow\quad 1\ H^+{}_{(aq)} \quad+\quad Cl^-{}_{(aq)}$$

$$1\ mol/L \qquad\qquad 1\ mol/L \qquad\qquad 1\ mol/L$$

**Step 2.** Use the concentration of $H^+_{(aq)}$ ions to find the concentration of $HCl_{(aq)}$.

According to the ionization equation in **Step 1, 1** mol/ $Cl^-_{(aq)}$ is produced from 1 mol/L $HCl_{(aq)}$. Knowing the concentration of $H^+_{(aq)}$, the concentration of $HCl_{(aq)}$ can be calculated according to the following equation:

$$[HCl_{(aq)}] = \frac{1\dfrac{mol\ HCl_{(aq)}}{L}}{1\dfrac{mol}{L}H^+} \times 0.100 \text{ mol/L } H^+_{(aq)}$$

$$= \textbf{0.100 mol/L } HCl_{(aq)}$$

**Step 3.** Since the concentration and volume of $HCl_{(aq)}$ are known, its number of moles can be found using the following equation:

$$\mathbf{n = C \times V} = 0.100 \text{ mol/L} \times 0.5000 \text{ L}$$

$$= \textbf{0.0500 mol } HCl_{(aq}$$

$$n = C \times V$$

**Step 4.** Convert amount of $HCl_{(aq)}$ from number of moles to amount in mass. To do this, the molar mass of $HCl_{(aq)}$ must first be calculated.

$$M = 1 \times 1.01 \text{ g/mol H } + 1 \times 35.45 \text{ g/mol Cl } = \textit{36.5 g/mol}$$

Using the formula, $\mathbf{m = n \times M}$ we get

$$\text{Mass} = 0.0500 \text{ mol} \times 36.5 \text{ g/mol} = \textbf{1.82 g } HCl_{(aq)}$$

$$m = n \times M$$

**Exercise 13.9 (p.225)**

## Finding Percentage concentration

### *Problem:*

**1.** Vinegar is sold as a 5.00 % V/V solution of pure acetic acid in water.

(a) What volume of acetic acid and water must be mixed to make 10.0 L of a 5.00 % V/V solution of vinegar?

*Provided:*

Volume of vinegar = 10.0 L

Concentration of vinegar = 5.00 %

*Required:*

The volumes of acetic acid and water to be added together.

*Steps to be taken:*

**Step 1.**

Using the formula, $C = \dfrac{V_{solute}}{V_{solution}} \times 100 \%$ and making $V_{solute}$ the subject of the equation we get

$$V_{solute} = C \times \frac{V_{solution}}{100\%} = \frac{5.00 \% \times 10.0 \text{ L}}{100 \%} = 0.500 \text{ L acetic acid}$$

**Step 2.** Find the volume water that must be added to make a total volume of 10.0 L.

$$= 10.0 \text{ L} - 0.500 \text{ L} = \textbf{9.50 L}$$

The mixture would have **9.50 L water** and **0.500 L acetic acid.**

a) If the density of acetic acid is **1.049 g/mL**, find the mass of the acetic acid used.

**Step 3.** Find the mass of the acid used.

$$0.500 \text{ L of the acid} = 5.00 \times 10^2 \text{ mL}$$

Using the formula for density, $\quad d = \dfrac{m}{V}$ and rearranging it we get for mass:

$$\text{mass} = d \times v = 1.049 \text{ g/mL} \times 5.00 \times 10^2 \text{ mL} = \textbf{525 g } \mathbf{C_2H_4O_2}$$

(a)  How many moles of acetic acid used does your answer in part (b) correspond to?

**Step 4.** To solve this problem, the molar mass of the acetic acid must first be calculated.

$$M = 2 \times 12.01 \text{ g/mol C} + 4 \times 1.01 \text{ g/mol H} + 2 \times 16.00 \text{ g/mol O} = \textbf{60.06 g/mol}$$

Using the formula, $n = \dfrac{m}{M}$  we get $n = \dfrac{525 \text{ g}}{60.06 \text{ g/mol}} = \textbf{8.74 mol } \mathbf{C_2H_4O_2}$

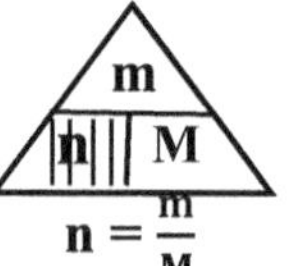


## *Problem:*

2. The concentration of sulphuric acid, $H_2SO_{4(aq)}$ in a sample of solution is found to be 1.20 mol/L. If the density of $H_2SO_{4(aq)}$ is 1.06 g/mL, calculate the volume percent of the sulphuric acid in the solution.

*Provided:*

Concentration of $H_2SO_{4(aq)}$ = 1.100 mol/L mol/L

Density of $H_2SO_{4(aq)}$ = 1.06 g/mL

*Required:*

The volume percent of the sulphuric acid in the solution

*Steps to be taken:*

**Step 1.** Convert the density of the acid from g/mL to g/L

$$d = 1.06 \text{ g/mL} \times 1000 \ mL/L = \textbf{1160 g/L}$$

**Step 2.** State the number of moles of $H_2SO_{4(aq)}$ in one litre of solution = 1.10 mol.

**Step 3.** Find the mass of $H_2SO_{4(aq)}$ in one litre of solution of solution. To do this, first find its the molar mass.

$$M = 2 \times 1.01 \text{ g/mol H} + 1 \times 32.07 \text{ g/mol S} + 4 \times 16.00 \ \text{g/mol O} = \textbf{\textit{98.07 g/mol}}$$

Using the formula, $\mathbf{m = n \times M}$ we get

$$\text{Mass} = 1.10 \text{ mol/} \times 98.07 \text{ g/mol} = \textbf{108.6 g}$$

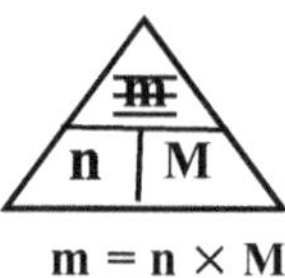


**Step 4.** Find the volume of $H_2SO_{4(aq)}$.

Since density $(d) = \dfrac{mass(m)}{volume(V)} = d = \dfrac{m}{V}, \ V = \dfrac{m}{d}$

Substituting values into $V = \dfrac{m}{d}$ we get $V = \dfrac{108.6 \text{ g}}{1160 \text{ g/L}} = 9.36 \times 10^{-2} \text{ L}$

**Step 5.** Find the volume percentage of $H_2SO_{4(aq)}$.

$$\dfrac{9.36 \times 10^{-2} \text{ L}}{1.00 \text{ L}} \times 100\% = 9.36 \text{ \% by volume}$$

**3.** A 100.0 mL box of mixed fruit juice has a fructose ($C_6H_{12}O_6$) concentration of 12.0 % mass-volume. What mass of fructose would be found in 1.00 L of fruit juice?

*Provided:*

Volume of juice = 100.0 mL

Concentration of juice = 12.0 %

*Required:*

The mass of fructose would be found in 1.00 L of fruit juice

*Steps to be taken:*

**Step 1.** Find the mass of fructose in a 100.0 mL box of juice, by coverting percentage composition to mass.

12.0 % M/V = 12.0 g/100 mL

**Step 2.** Find the mass of fructose in 1.00L (1000.0 mL/L) of juice

$$\frac{12.0 \text{ g}}{100.0 \text{ mL}} \times 1000 \text{ mL/L} = \mathbf{1.20 \times 10^2 \text{ g/L fructose}}$$

*Problem:*

**4.** Coffee brewed with pure water contains approximately 0.045 % M/V V concentration of caffeine. What mass of caffeine will be found in a 200-mL cup of unsweetened coffee?

*Provided:*

Concentration of caffeine in coffee: 0.045 % M/V

Volume of coffee = 200.0 mL

*Required:*

The mass of caffeine that would be found in a 200.0 mL cup of unsweetened coffee.

*Steps to be taken:*

**Step 1.** Find the mass of caffeinin a 100.0 mL cup, by coverting percentage composition to mass.

0.045 % M/V = 0.045 g/100.0 mL

**Step 2.** Find the mass of caffeine in 200.0 mL cup of tea

$$\frac{0.045 \text{ g}}{100.0 \text{ mL}} \times 200.0 \text{ mL} = \mathbf{0.090 \text{ g} = 90 \text{ mg of caffeine}}$$

*Problem:*

**5.** Plumber's solder is 33.0% tin and 67.0% lead M/M percent. How many moles of tin are present in a 500.0 g sample of solder?

*Provided:*

Composition of the solder: 33.0% tin and 67.0% lead M/M

*Required:*

The number of moles of tin present in a 500.0 g sample of solder

*Steps to be taken:*

**Step 1**. Find the mass of tin present in a 100.0 g sample of solder.

Since tin is 33.0 % M/M, the mass in a 100.0 g sample of solder will be 33.0 g

**Step 2**. Find the mass of tin present in a 500.0 g sample of solder

$$\frac{33.0 \text{ g}}{100.0 \text{ g}} \times 500.0 \text{ g} = \textbf{165 g tin}$$

**Step 3.** Find the number of moles of tin present in a 165 g tin.

Using the formula, $\mathbf{n} = \dfrac{\mathbf{m}}{\mathbf{M}}$ we get

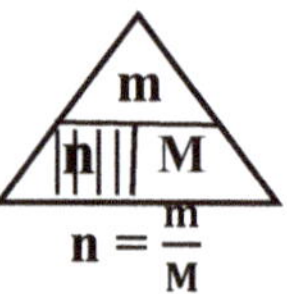


$$n = \frac{165 \text{ g}}{118.71 \text{ g/mol}} = \textbf{1.39 mol Sn}$$

## Exercise 13.11 (p.226)

**Very Low Concentrations**

*Problem:*

1. In an attempt to investigate the bioaccumulation of DDT up the food chain in a particular **lake with a DDT concentration of 5.00 ppb**, the tissues of some birds that feed on fish from the lake were analyzed. If 1.00 kg of bird tissues contained $3.00 \times 10^{-2}$ g of DDT, what is its concentration in pbb?

*Provided:*

1.00 kg of bird tissue (solvent) = $1.00 \times 10^{3}$ g

Mass of DDT (solute) in 1.00 kg of tissue = $3.00 \times 10^{-2}$ g

*Required:*

The concentration of DDT, in parts per billion in bird's tissue.

*Steps to be taken:*

**Step 1.** Using the formula, $\text{pbb} = \dfrac{\text{Mass of solute}}{\text{Mass of tissue}} \times 10^{9}$ we get,

$$\text{pbb} = \frac{3.00 \times 10^{-2} \text{ g}}{1.00 \times 10^{3} \text{ g}} \times 10^{9} = 3.000 \times 10^{3}$$

*Problem:*

2. If 1.0 kg of polluted air contains 0.015 g of carbon monoxide, what is its concentration in ppm?

*Provided:*

1.0 kg of polluted air (solvent) = $1.0 \times 10^{3}$ g

0.015 g of carbon monoxide (solute)

*Required:*

The concentration of carbon monoxide concentration in ppm?

*Steps to be taken:*

**Step 1.** Using the formula, $ppm = \dfrac{\text{Mass of solute}}{\text{Mass of solvent}} \times 10^6$ we get,

$$ppm = \dfrac{0.015 \text{ g}}{1.0 \times 10^3 \text{ g}} \times 10^6 = 15$$

*Problem:*

3. If the concentration of carbon dioxide in an early morning outdoor air is 350 ppm, what mass of it will be in 2.0 kg of air?

*Required:*

The mass of carbon dioxide (solute) in 2.0 kg of air

*Provided:*

Concentration of carbon onoxide in air = 350 ppm

Mass of air (solvent) = 2.0 kg = $2.0 \times 10^3$ g

*Steps to be taken:*

**Step 1.** Find the mass of 1 part carbon dioxide in air.

$$350 \text{ g} \div 10^6 \text{ parts} = 3.5 \times 10^{-4} \text{ g/parts}$$

**Step 2.** Since $2.0 \times 10^3$ g of air = $2.0 \times 10^3$ parts, the mass of $CO_2$ in it will be:

$$3.5 \times 10^{-4} \text{ g/part} \times 2.0 \times 10^3 \text{ parts} = \textbf{0.70 g carbon dioxide}$$

**Exercise 13.12 (p.227)**

**Finding Molar Concentration, given Density and Percentage by Volume**

*Problem:*

1. A solution of $H_2SO_{4(aq)}$ is 83.0% V/V and has a density of 1.22 g/mL at $20^\circ$ C, what is its molar concentration?

*Required:*

The molar concentration of $H_2SO_{4(aq)}$

*Provided:*

V/V concentration of $H_2SO_{4(aq)}$ = 83.0%

Density of $H_2SO_{4(l)}$ = 1.22 g/mL at $20^\circ$ C

Formula of the acid = $H_2SO_{4(aq)}$

*Steps to be taken:*

**Step 1.** Find the volume of pure $H_2SO_{4(l)}$ that was used to make the solution.

Since the concentration is 83.0% V/V, its volume = 83.0 mL/ 100 mL of solution

**Step 2.** Find the mass of $H_2SO_{4(l)}$ in 83.0 mL of density 1.22 g/mL at $20^\circ$ C

Using the formula, $Density = \dfrac{\text{Mass}}{\text{Volume}}$, and making mass the subject of the equation we get mass = density $\times$ volume = 1.22 g/mL $\times$ 83.0 mL = 101 g

**Step 3.** Find the number of moles of $H_2SO_{4(l)}$ in 101 g. To do this, first find its molar mass.

$$M = 2 \text{ mol} \times 1.01 \text{ g/mol H} + 1 \text{ mol} \times 32.07 \text{ g/mol S} + 4 \text{ mol} \times 16.00 \text{ g/mol} = \textbf{\textit{98.09 g/mol}}$$

sing the formula, $n = \dfrac{m}{M}$ we get

$$n = \frac{101 \text{ g}}{98.09 \text{ g/mol}} = \textbf{1.03 mol } H_2SO_{4(l)}$$

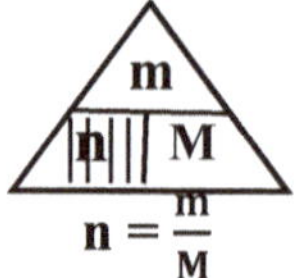

**Step 4.** Find the molar concentration of $H_2SO_{4(aq)}$

The total volume of the acid is 100.0 mL = 0.1000 L

Using the formula, $C = \dfrac{n}{V}$ we get, $\dfrac{1.03 \text{ mol}}{0.1000 \text{ L}} = \textbf{10.3 mol/L } H_2SO_{4(aq)}$

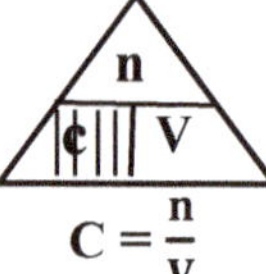

***Problem:***

**Analysis 1. (p.232)**

## Solubility of gases in water

**1.** Carbon dioxide       **2.** Carbon dioxide       **3.** No

**4.** Gases dissolve less in warm solvents than in cold solvents. In this case, as the soda was heated up by the hot water, the dissolved carbon dioxide gas molecules gained more kinetic energy and escaped at a greater rate from the soda, than it did in the cooler beaker. This was evident by the vigorous effervescence produced. As the carbon dioxide gas accumulated in the beaker, it pushed the air out of it. When the level of carbon dioxide gas reached the tip of the candle it extinguished it, since carbon dioxide does not support combustion. This clearly shows that gases are less soluble in warm water than in colder water.

**5.** The effects of thermal pollution in bodies of water would be due to the lack of dissolved gases such as oxygen and carbon dioxide in the water. Less dissolved carbon dioxide would severely affect production of producers (aquatic algae) by photosynthesis, and this can be detrimental to life forms in aquatic ecosystems; production and productivity in these bodies of water will be asversely affected. The same can be implied about the lack of dissolved oxygen, which is essential for cellular respiration.

**6.** In the warm can of soda less gas dissolved in the solution because of the higher temperature so more gas particles occupy the space above the solution creating a higher pressure there than in the cold can. Because of this higher pressure, the gas made a louder sound as it tried to escape than it did in the cold can.

**7.** Based on the observations, it can be concluded that less gases dissolve in warm aqueous solutions than in colder ones.

**Analysis 2. (p.234)**

## Solubility curves

**1.** Unsaturated       **2.** Saturated

**3.** As the temperature of the solution increases the solubility of $KNO_3$ in it increases.

**Conclusion:** The solubility of $KNO_3$ increases with temperature rise.

**Analysis 3. (p.235)**

**1.** KI       **2.** 72.0°C       **3.** 45 g       **4.** $SO_{2(g)}$ and $NH_{3(g)}$

**5.** $NH_3$ molecules are attracted to water molecules by both hydrogen bonding and London dispersion forces, where as $SO_2$ molecules which are mildly polar, are attracted to water molecules only by dipole-dipole forces. Ammonia is thus more soluble in water than sulfur dioxide because ammonia molecules are attracted to water molecules by stronger intermolecular forces than do the molecules of sulfur dioxide.

**6.** 73.0 g.

**7.** No, because the exact volume of the solution is unknown; different amounts of solutes were dissolved in 100.0 mL of water; the volume of each the solutions is what resulted from the addition of solute to 100.0 mL of water. The only way of calculating their molar concentrations is for the volume of each of the solutions to be measured accurately and use the masses of solutes dissolved, as provided in the graphs.



1.Lowering temperatures cause less solubility and therefore eventual crystallization occurs.

2.Gently heat the jar in a water bath to a higher temperature but less than 40 °C, until all the crystals just disappear. Do not over-heat, because a higher temperature will destroy some of the beneficial properties of honey.

3.Particles of wax and pollens in raw honey act as nuclei for crystallization; these are only present in miniscule amounts in treated honey, so crystallization is much slower in these.

4.Nothing changes except that the sugars have crystallized out of solution.

5.Fructose could be found in the liquid part since it has a higher solubility than glucose. Since glucose has a lower solubility, it crystallizes out of solution before fructose does.

## CHAPTER 15:  Soap and Detergents

**Part A:**

Table 15.1 (p.241)

| Water samples | Observation After Shaking |
|---|---|
| **1.** Distilled water + soap | A lot of lather was readily formed |
| **2.** Distilled water + soap + $MgCl_{2(aq)}$ | A creamy precipitate was formed. Only a small amount of lather was formed |
| **3.** Distilled water + soap + $CaCl_{2(aq)}$ | A creamy precipitate was formed. Only a small amount of lather was formed |
| **4.** Distilled water + soap + $NaCl_{(aq)}$ | A lot of lather was readily formed |



1.Test tubes 1 and 4

2.Test tubes 2 and 4.

3.Test tubes 2 and 4. Precipitates were formed because the $Mg^{2+}_{(aq)}$ and $Ca^{2+}_{(aq)}$ ions present reacted with the soap molecules to produce compounds of magnesium and calcium that are not soluble in water.

4. test tubes 2 and 4

## Conclusion

1.  $Mg^{2}_{(aq)}$ and $Ca^{2+}_{(aq)}$ ions are responsible for the hardness of water.

$$Mg^{2+}_{(aq)} + 2\ C_{17}H_{35}COO^{-}_{(aq)} \longrightarrow Mg(C_{17}H_{35}COO)_{2(s)}$$

$$Ca^{2+}_{(aq)} + 2\ C_{17}H_{35}COO^{-}_{(aq)} \longrightarrow Ca(C_{17}H_{35}COO)_{2(s)}$$

**Part B:**

**Table15.2**

| | Water samples | Observation Before Shaking |
|---|---|---|
| **1** | Distilled water <br> + $Na_2CO_{3(aq)}$ | No visible change took place |
| **2** | Distilled water + $Na_2CO_{3(aq)}$ <br> + $MgCl_{2(aq)}$ | A creamy precipitate was formed |
| **3** | Distilled water + $Na_2CO_{3(aq)}$ <br> + $CaCl_{2(aq)}$ | A creamy precipitate was formed |
| **4** | Distilled water + $Na_2CO_{3(aq)}$ <br> + $NaCl_{(aq)}$ | No visible change took place |

## Analysis 2. (p.242)

1.(a) Test tubes 2 and 3.

(b) $Mg^{2+}_{(aq)} + CO_3^{2-}_{(aq)} \rightarrow MgCO_{3(s)}$

$Ca^{2+}_{(aq)} + CO_3^{2-}_{(aq)} \rightarrow CaCO_{3(s)}$

**Table15.3**

| Water samples | Observation After Shaking |
|---|---|
| **1.** Distilled water <br> + soap + $Na_2CO_{3(aq)}$ | A lot of lather was readily formed |
| **2.** Distilled water + $Na_2CO_{3(aq)}$ <br> + soap + $MgCl_{2(aq)}$ | A lot of lather was readily formed <br> A precipitate remains |
| **3.** Distilled water + $Na_2CO_{3(aq)}$ <br> + soap + $CaCl_{2(aq)}$ | A lot of lather was readily formed <br> A precipitate remains |
| **4.** Distilled water + $Na_2CO_{3(aq)}$ <br> + soap + $NaCl_{(aq)}$ | A lot of lather was readily formed |

## Analysis 3. (p.243)

1. Yes. Test tubes 2 and 4 produced significantly more lather than in procedure A.

2. No

3. The water was softened because the $CO_3^{2-}_{(aq)}$ ions removed the $Mg^{2+}_{(aq)}$ and $Ca^{2+}_{(aq)}$ ions from the water in the form of precipitates. When they are in the form of precipitates, they cannot affect the soap.

**Table15.4**

| Water samples | Observation After Shaking |
|---|---|
| 1.      Distilled water + detergent | A lot of lather was readily formed |
| 2. Distilled water + detergent + $MgCl_{2(aq)}$ | No precipitate but lots of lather formed |
| 3. Distilled water + detergent + $CaCl_{2(aq)}$ | No precipitate but lots of lather formed |
| 4. Distilled water + detergent + $NaCl_{(aq)}$ | A lot of lather was readily formed |

**Analysis 4. (p.244)**

**1** No      **2.** No      **3.** No

**Conclusion:**

Detergents work well in hard water because its molecules do not form precipitates with $Mg^{2+}_{(aq)}$ and $Ca^{2+}_{(aq)}$ ions; they are thus able to cleanse without any interferences from these ions.

# Chapter Review:  Solution and solubility (Chapter 12-15)

## Matching

**Match each term in the table below with the correct statements that follow**

| | | | |
|---|---|---|---|
| A | Water | G | Solvent |
| B | Solute | H | mol/L |
| C | Polar | I | Hydrogen bonds |
| D | Ionization | J | Saturated |
| E | Water of crystallization | K | Volumetric flask |
| F | Inter molecular | L | Electrostatic |

| | |
|---|---|
| 1. | This is used to dissolve the solute during the solution process. |
| 2. | This is called a universal solvent. |
| 3. | These hold the water molecules together in its liquid form. |
| 4. | This is dissolved by the solvent during the solution process. |
| 5. | The unit used to describe molar concentration. |
| 6. | These types of molecules attract each other by dipole-dipole forces. |
| 7. | The process by which a molecule breaks up into ions. |
| 8. | A solution that cannot dissolve any more solute at a specific temperature. |
| 9. | Hydrated salts have these. |
| 10. | Equipment used to accurately prepare solutions. |
| 11. | Force of attraction between molecules. |
| 12. | Force of attraction between the ions in an ionic compound. |

## True or False

**Read each of the following statements and then decide if it is *True* or *False*.**

| | |
|---|---|
| 1. | Water is a bent polar molecule. |
| 2. | Grease can be dissolved by water. |
| 3. | The mixing of water and a concentrated acid can be highly exothermic. |
| 4. | Evaporating water from a solution changes its concentration. |
| 5. | Adding water to a solution decreases the solution's concentration. |
| 6. | In any solution of $ZnCl_{2(aq)}$, the concentration of $Zn^{2+}_{(aq)}$ and $Cl^-_{(aq)}$ will always be the same. |
| 7. | Hydrogen bonding is the major intermolecular force of attraction between water molecules. |
| 8. | When a negative ion is solvated by water molecules, it is surrounded by the oxygen ends of the water molecules. |
| 9. | The solvation of ions is by water molecules is an endothermic process. |

| 10. | Water molecules auto ionize to produce $H^+_{(aq)}$ and $O^{2-}_{(aq)}$ ion. |
| 11. | The process whereby the ions of an ionic compound separate is called ionization. |
| 12. | The concentration of $NO_3^-{}_{(aq)}$ in 0.1 mol/L $KNO_{3(aq)}$ is half of that present in 0.1 mol/L $Ba(NO_3)_{2(aq)}$. |
| 13. | The concentration of $Cl^-_{(aq)}$ in 0.2 mol/L $NaCl_{(aq)}$ is the same as in 0.1 mol/L $BaCl_{2(aq)}$. |
| 14 | If a solution of salt is heated for some time, the amount of salt in it increases. |

## Multiple Choice

**Choose the letter that best answers the question.**

1. Which of the following is **not** a solution?

| | | | | |
|---|---|---|---|---|
| a | carbon dioxide in water | d. | salt in water |
| b | copper in gold | e. | alcohol in water |
| c | sand in water | | |

2. Water molecules are attracted to each other by

| | | | | |
|---|---|---|---|---|
| a | London dispersion forces | d. | hydrogen bonds |
| b | ionic bonds | e. | both a and d |
| c | magnetic forces | | |

3. Which of the following processes are involved in dissolving an ionic compound?

| | | | | |
|---|---|---|---|---|
| a | breaking of the hydrogen bonds in water | d. | none of the above |
| b | breaking of the ionic bonds | e. | a, b and c |
| c | solvation of the ions | | |

4. Water dissolves ethyl alcohol because they are both

| | | | | |
|---|---|---|---|---|
| a | colourless molecules | d. | non-polar molecules |
| b | small molecules | e. | none of the above |
| c | polar molecules | | |

5. In aqueous solutions, the solvent is always

| | | | | |
|---|---|---|---|---|
| a | air | d. | oil |
| b | alcohol | e. | none of the above |
| c | water | | |

6. A solution that has a relatively large quantity of solute dissolved in it is said to be

| | | | |
|---|---|---|---|
| *a* | concentrated | *d.* | strong |
| *b* | dilute | *e.* | a and c only |
| *c* | saturated | | |

7. Which of the following factors determine the concentration of a solution?

| | | | |
|---|---|---|---|
| *a* | the volume of solvent | *d* | the size of the particles |
| *b* | the temperature of the solution | *e* | a and c only |
| *c* | the number of moles particles dissolved | | |

8. The mass of sodium hydroxide dissolved in 500.0 mL of 0.20 mol/L $NaOH_{(aq)}$ is

| | | | |
|---|---|---|---|
| *a* | 4.0 g | *d* | 1.0 g |
| *b* | 0.20 g | *e* | none of the above |
| *c* | 2.0 g | | |

9. A solution having one mole of a solute dissolved in 500.0 mL of solution will have a concentration of

| | | | |
|---|---|---|---|
| *a* | 1.00 mol/L | *d* | 0.250 mol/L |
| *b* | 2.00 mol/L | *e* | none of the above |
| *c* | 0.500 mol/L | | |

10. What volume of water should be added to 250 mL of a 2.0 mol/L $HCl_{(aq)}$ solution to dilute it to 0.5 mol/L?

| | | | |
|---|---|---|---|
| *a* | 500 mL | *d* | 750 mL |
| *b* | 1.0 L | *e* | 375 mL |
| *c* | 250 mL | | |

11. If 4.00 g of $NaOH_{(s)}$ is dissolved in water to make 200.0 mL of solution, then its concentration will be

| | | | |
|---|---|---|---|
| *a* | 2.00 mol/L | *d* | 0.500 mol/L |
| *b* | 4.00 mol/L | *e* | none of the above |
| *c* | 1.00 mol/L | | |

12. Barium hydroxide dissociates in water as follows: $Ba(OH)_{2(aq)} \rightarrow Ba^{2+}_{(aq)} + 2\ OH^-_{(aq)}$. If the concentration of the $Ba(OH)_{2(aq)}$ is 0.50 mol/L, the $[Ba^{2+}_{(aq)}]$ and $[OH^-_{(aq)}]$ will be

| | | | |
|---|---|---|---|
| *a* | $[Ba^{2+}_{(aq)}] = 0.50$ mol/L | *d* | $[OH^-_{(aq)}] = 1.0$ mol/L |
| *b* | $[OH^-_{(aq)}] = 0.50$ mol/L | *e* | only a and d |
| *c* | $[Ba^{2+}_{(aq)}] = 1.0$ mol/L | | |

13. If 9.52 g of $MgCl_{2(s)}$ is dissolved in water to make 400.0 mL of solution, the $[Cl^-_{(aq)}]$ will be

| | | | | |
|---|---|---|---|---|
| a | 0.500 mol/L | d | 4.00 mol/L |
| b | 1.00 mol/L | e | none of the above |
| c | 2.00 mol/L | | |

14. Hardness of water is caused by what ions?

| | | | | |
|---|---|---|---|---|
| a | $Na^+$ and $K^+$ | d | $Mg^{2+}$ and $Al^{3+}$ |
| b | $Ca^{2+}$ and $Mg^{2+}$ | e | none of the above |
| c | $Cl^-$ and $I^-$ | | |

15. Copper(II) sulphate pentahydrate dissociates in water as follows:

$CuSO_4.5H_2O_{(s)} + H_2O_{(l)} \rightarrow Cu^{2+}_{(aq)} + SO_4^{2-}_{(aq)}$. What mass of $CuSO_4.5H_2O_{(s)}$ is needed to prepare 250.0 mL of 0.20 mol/L of $CuSO_{4(aq)}$?

| | | | | |
|---|---|---|---|---|
| a | 24.96 g | d | 15.96 g |
| b | 12.48 | e | none of the above |
| c | 7.98 g | | |

16. In what volume of solution must 11.10 g of $CaCl_{2(s)}$ be dissolved in to produce 0.20 mol/L $Cl^-_{(aq)}$?

| | | | | |
|---|---|---|---|---|
| a | 2.0 L | d | 1.0 L |
| b | 0.50 L | e | none of the above |
| c | 4.0 L | | |

17. Water does not dissolve grease because

| | | | | |
|---|---|---|---|---|
| a | they are colourless molecules | d | they are both non-polar molecules |
| b | they are small molecules | e | water is polar but grease is non-polar |
| c | they are both polar molecules | | |

18. It is dangerous to dump untreated sewage and other effluent in lakes and rivers because

| | | | | |
|---|---|---|---|---|
| a | they contain poisons | d | all of the above |
| b | they can cause diseases | e | none of the reasons given |
| c | they can cause algal blooms | | |

**Chapter Review: Solution (12-15) Solution:**

**Matching:**

1.) G    2.) A    3.) I    4.) B    5.) H    6.) C    7.) D    8.) J

9.) E    10.) K    11.) F    12.) L

**True/False**

1.) T    2.) F    3.) T    4.) T    5.) T    6.) F    7.) T    8.) F

9.) F    10.) F    11.) F    12.) T    13.) T    14.) F

**Multiple Choice**

1.) C    2.) E    3.) E    4.) C    5.) C    6.) A    7.) E    8.) A

9.) B    10.) D    11.) D    12.) E    13.) A    14.) B    15.) B    16.) B

17.) E    18.) D

## Exercise 16.1 (p.252-253)

1. **a)** The presence of too much E. Coli bacteria means that water is not well sterilized. This can be rectified by doing more effective sanitizing of the water.
High nitrates means that there was not enough time spent by the effluent in the anoxic zone for nitrifying bacteria to act. To solve this problem, allow the effluent to spend more time in the anoxic zone to ensure all the nitrates are decomposed by the nitrifying bacteria.
   **b)** If there is an inadequate supply of oxygen in the oxic zone, some nitrogenous materials such as urea and ammonia might escape oxidation to nitrates. The water ensuing water will have unchanged ammonia and urea.
   **c)** Because the flow of water is considerably slowed down in the primary clarifier, if it stays too long there, microbial activities will use up any dissolved oxygen. It will therefore cost more to replenish the lost oxygen by the pumping process in the oxic zone.
   d) There might not be enough bugs for the breakdown of the sewage and lots of undecomposed materials can move along in the water.

**1.  Site A**
Water at sight A with high turbidity, *E. Coli*, nitrates and ammonia concentrations, couple with low amounts of dissolved oxygen, is typical of those from sources with untreated sewage. Such sources could be runoff from feeding lots, pit latrines etc. The low dissolved oxygen is characteristic of the high BOD of the water.

**Site B**
Water at site B is most likely from uncontaminated sources such as water exiting a sewage treatment plants or spring water. In these, because of the absence of raw sewage, there is low turbidity with low *E. Coli*, nitrates and ammonia concentrations. The high dissolved oxygen is indicative of the absence of BOD.

2. a) Water from sample A is unsafe to drink because the levels of lead, nitrates, barium and E.Coli ,far exceed the safe limits set by the Canadian guideline for drinking water quality. Water from sample B is safe to drink because all of the above-mentioned substances are within the allowed ranges.

## Analysis 4.  (p.259)

1. a) Test tubes 7 and 2
   b) Test tubes 10 and 5
   c)  Test tubes 7 and 2
   d) Test tubes 10 and 5
   e) It was important to stopper all the test tube to prevent any dissolved gases in the test tubes ($O_{2(aq)}$ and $CO_{2(aq)}$) from escaping and to prevent gases from the air to enter. If these happened, it would have altered the results.

**Conclusion:**
As the amount of organic matter in the water increased, there was a proportionate decrease in the amount of dissolved oxygen.

**Additional questions:**

2.  a) The purpose of test tubes 1 and 6 was to act as controls; nothing, except the indicators were placed in these. The colors of the indicators in these were allowed to remain unchanged so they were used to monitor the microbial activities in the other test tubes by comparing the colour changes that took place in them.

   **b)** All of the dissolved oxygen will be used up and the carbon dioxide level will be high.   Also, anoxic condition will arise with the proliferation of anaerobic microbes.

   **c)** The changes started from the bottom of the test tubes because that was where the microbial activities were happening; there was no mixing of the contents in the test tubes.

**d)** There is a direct correlation between the amount of oxygen used and the amount of carbon dioxide produced. As the intensity of the yellow color in the test tubes in figure 16.3 (b) increases; which is indicative of increased carbon dioxide production, there is corresponding decrease in the intensity of the blue color in the test tubes in figure 16.3 (a); which is indicative of oxygen consumption.

During cellular respiration, oxygen is consumed, and carbon dioxide is liberated. The oxygen consumed is extracted from that dissolved in the water; this removal of oxygen from water is what caused the methylene blue indicator to go colorless. The carbon dioxide, which was liberated, dissolved in the water, and this is what caused the bromothymol indicator to turn yellow.

**e)** If all the stoppers were removed from the test tubes, it would allow for free diffusion of gases from the test tubes into the air, and from the air into the test tubes. For the test tubes in figure 16.3 (a), oxygen from the air will diffuse into them and the blue color of the indicator (methylene blue) will return throughout the test tubes. For the test tubes in figure 16.3 (b), dissolved carbon dioxide from the test tubes will diffuse from them into the air and the blue color of the indicator (bromotymol blue) will return throughout the test tubes.

**f)** As the amount of organic matter increased, the number of bacteria increased. This is evident by the change in the intensity of color changes in the test tubes as the amount of rice was increased.

**3.** Every time biodegradable organic wastes are dumped into a body of water the ensuing microbial activities depletes the dissolved oxygen. Oxygen is essential for cellular respiration inaquatic organisms. If people do not refrain from this practice, it may lead to profound oxygen depravation, resulting in death of aquatic plants and animal. Bodies of water may even become anoxic.

**4.** Spill weirs, along the course of rivers polluted with organic wastes, allow for greater mixing of air with the water. As this happens, more oxygen gets dissolved which facilitates the decomposition of the organic wastes by the microbes present in the water.

**6.** Pumping air into fish tanks replenishes the oxygen consumed by the fish and provides oxygen to the microbes to decompose any organic wastes produced by the fish. If they discontinue this practice, the fish as well the decomposers will be deprived of oxygen which can lead to fish death and anoxic conditions.

# Chapter Review:  Sewage treatment (Chapter 16) solution

## Matching

*Match the item in the table below with the statements that follows*

| | | | |
|---|---|---|---|
| A | Sewage | G | Aerobic |
| B | Oxic zone | H | Anaerobic |
| C | Anoxic zone | I | Algal bloom |
| D | BOD | J | Eutrophic |
| E | Bugs | K | Biodegradable |
| F | Activated sludge | L | MAC |

| | |
|---|---|
| 1. | A condition that results when dissolved oxygen in a body of water is severely reduced. |
| 2. | A mixture of faeces and wastewater from domestic and industrial use. |
| 3. | In sewage treatment, this intermediate waste consists of organic material wastes and living   microbes. |
| 4. | These substances can be broken down by bacteria and fungi. |
| 5. | This term is used to collectively describe all the microbes utilized to feed on sewage. |
| 6. | The maximum acceptable concentration of any substance permitted to be in drinking water. |
| 7. | These types of microbes use dissolved oxygen to feed. |
| 8. | These types of microbes extract the oxygen bonded in nitrate ions to feed. |
| 9. | The zone in the biological treatment basin that is well aerated. |
| 10. | The zone in the biological treatment basin that is deprived of oxygen. |
| 11. | A measure of the amount of dissolved oxygen needed by microbes to feed on organic carbon. |
| 12. | This type of condition is normally produced when lots of fertilizer enters bodies of water such as rivers and lakes. |

## True or False

**Read each of the following statements and then decide if it is *True* or *False*.**

| | |
|---|---|
| 1. | Too much fertilizer in water cannot affect the amount of dissolved oxygen. |
| 2 | Too much sodium in drinking water is not healthy for hypertensive people. |
| 3 | Raw sewage is laced with a host of pathogens. |
| 4 | All inorganic substances are removed from water after treatment. |
| 5 | Sewage consists mainly of water. |
| 6 | Acid rain raises the pH of water. |
| 7 | The use of UV is less eco-friendly than chlorine for water treatment. |

| 8 | Coagulants allow colloidal materials to clump together and settle. |
|---|---|
| 9 | A fish tank operating without an air pump could become anoxic. |
| 10 | Turbid water has less dissolved solids than transparent water. |

## Multiple Choice

**Choose the letter that best answers the questions.**

1. Which of the following are examples of coagulants?

| | | | | |
|---|---|---|---|---|
| *a* | Zinc Chloride | *d* | Ferric sulphate |
| *b* | Aluminum sulphate | *e* | b and c |
| *c* | Calcium Nitrate | | |

2. In the absence of oxygen methylene blue goes

| | | | | |
|---|---|---|---|---|
| *a* | blue | *d* | colourless |
| *b* | yellow | *e* | to none of the above |
| *c* | pink | | |

3. Which of the following substances are non-biodegradable?

| | | | | |
|---|---|---|---|---|
| *a* | plastics | *d* | oil spills |
| *b* | metals | *e* | all except c |
| *c* | wood | | |

4. In the presence of lot of carbon dioxide bromothymol blue goes

| | | | | |
|---|---|---|---|---|
| *a* | blue | *d* | colourless |
| *b* | yellow | *e* | to none of the above |
| *c* | pink | | |

5. Which of the following substances are biodegradable?

| | | | | |
|---|---|---|---|---|
| *a* | Fruits | *d* | Organic wastes |
| *b* | Fallen leaves | *e* | All of the above |
| *c* | Dead animals | | |

6. Which of the following are examples of heavy metals found in sewage?

| | | | | |
|---|---|---|---|---|
| *a* | Lead | *d* | Cadmium |
| *b* | Mercury | *e* | All of the above |
| *c* | Copper | | |

7. Drinking contaminated water can cause

| | | | |
|---|---|---|---|
| *a* | cholera | *d* | hepatitis A |
| *b* | dysentery | *e* | all of the above |
| *c* | guinea worms | | |

8. Which of the following would **not** affect the amount of dissolved oxygen?

| | | | |
|---|---|---|---|
| *a* | rise in temperature | *d* | fertilizers |
| *b* | organic wastes | *e* | windy conditions |
| *c* | arsenic | | |

9. Dumping of Lye in water

| | | | |
|---|---|---|---|
| *a* | reduces dissolved oxygen | *d* | lowers the pH |
| *b* | increases dissolved carbon dioxide | *e* | has no effect |
| *c* | raises the pH | | |

10. An increase in the turbidity of water

| | | | |
|---|---|---|---|
| *a* | lowers the rate of photosynthesis | *d* | changes pH |
| *b* | increases transparency | *e* | a and d |
| *c* | decreases transparency | | |

11. Which of the following incidents was associated with the *E. Coli* bacteria?

| | | | |
|---|---|---|---|
| *a* | Flint drinking water | *d* | The Chernobyl disaster |
| *b* | OMAI gold mine | *e* | None of the above |
| *c* | Walkerton drinking water | | |

12. Water contaminated with heavy metals such as mercury enter our bodies

| | | | |
|---|---|---|---|
| *a* | by drinking the contaminated water | *d* | by inhalation |
| *b* | ingestion via food chains | *e* | a and b |
| *c* | by bathing in the water | | |

13. Aluminum sulphate and ferric sulphate are used to purify water because they

| | | | |
|---|---|---|---|
| *a* | kill pathogens | *d* | regulate the pH |
| *b* | remove odours | *e* | do all of the above |
| *c* | coagulate colloidal materials | | |

14.

| | Allowing polluted water in rivers to flow along spill weirs facilitates purification by | | |
|---|---|---|---|
| *a* | allowing the water to flow fast | *d* | allowing dissolve solids to settle |
| *b* | allowing more aeration to take place | *e* | none of the above |
| *c* | allowing more photosynthesis to happen | | |

15.

| | Activated charcoal is used in filters to purify water by removing | | |
|---|---|---|---|
| *a* | carbon dioxide | *d* | heavy metals |
| *b* | odours | *e* | acidity |
| *c* | pathogens | | |

## Chapter Review: Sewage Treatment (Chapter 16) Solutions

**Matching:**

| 1.) J | 2.) A | 3.) F | 4.) K | 5.) E | 6.) L | 7.) G | 8.) H | 9.) B | 10.) C |
|---|---|---|---|---|---|---|---|---|---|

| 11.) D | 12.) J |
|---|---|

**True/False**

| 1.) F | 2.) T | 3.) T | 4.) F | 5.) T | 6.) F | 7.) F | 8.) T | 9.) T | 10.) F |
|---|---|---|---|---|---|---|---|---|---|

**Multiple Choice**

| 1.) E | 2.) D | 3.) E | 4.) B | 5.) E | 6.) E | 7.) E | 8.) D | 9.) C | 10.) E |
|---|---|---|---|---|---|---|---|---|---|

| 11.) C | 12.) E | 13.) C | 14.) B | 15.) E |
|---|---|---|---|---|

# Chapter 17: Acids and bases



Use the Brönsted-Lowry theory to identify the conjugate acid-base pairs in the following acid-base reactions.

(a) $HCO_3^-{}_{(aq)} + S^{2-}{}_{(aq)} \rightleftharpoons HS^-{}_{(aq)} + CO_3^{2-}{}_{(aq)}$

(b) $CO_3^{2-}{}_{(aq)} + HC_2H_3O_2 \rightleftharpoons HCO_3^-{}_{(aq)} + C_2H_3O_2^-{}_{(aq)}$

(c) $H_3PO_4{}_{(aq)} + OCl^-{}_{(aq)} \rightleftharpoons HClO_{(aq)} + H_2PO_4^-{}_{(aq)}$

(d) $HSO_4^-{}_{(aq)} + HPO_4^{2-}{}_{(aq)} \rightleftharpoons SO_4^{2-}{}_{(aq)} + H_2PO_4^-{}_{(aq)}$

**Answers:**
a) $HCO_3^-{}_{(aq)}$ and $CO_3^{2-}{}_{(aq)}$,     $HS^-{}_{(aq)}$ and $S^{2-}{}_{(aq)}$
b) $HC_2H_3O_2$ and $C_2H_3O_2^-{}_{(aq)}$,   $HCO_3^-{}_{(aq)}$ and $CO_3^{2-}{}_{(aq)}$
c) $H_3PO_4{}_{(aq)}$ and $H_2PO_4^-{}_{(aq)}$,   $HClO_{(aq)}$ and $OCl^-{}_{(aq)}$
d) $HSO_4^-{}_{(aq)}$ and $SO_4^{2-}{}_{(aq)}$,   $H_2PO_4^-{}_{(aq)}$ and $HPO_4^{2-}{}_{(aq)}$



**1.** a) A weak acid because it conducts electricity poorly.
b) Below 7 but close to 7.
c) $HC_6H_7O_6{}_{(aq)} + H_2O_{(l)} \rightleftharpoons C_6H_7O_6^-{}_{(aq)} + H_3O^+{}_{(aq)}$
d) $H_3O^+{}_{(aq)}$ and $H_2O_{(l)}$) and ($HC_6H_7O_6{}_{(aq)}$ and $C_6H_7O_6^-{}_{(aq)}$)
e) 1 mole
$$HC_6H_7O_6{}_{(aq)} + NaOH \rightarrow NaC_6H_7O_6 + H_2O_{(l)}$$

**2.** a) A strong base because it ionises completely. There is a predominance of $OH^-{}_{(aq)}$ ions.
b) The predominance of $K^+{}_{(aq)}$ and $OH^-{}_{(aq)}$ allows for excellent conduction.
c) Well above 7
d) i) $KOH + H_2SO_4{}_{(aq)} \rightarrow KHSO_4{}_{(aq)} + H_2O_{(l)}$
ii) The solution is acidic because the $KHSO_4{}_{(aq)}$ is an acid salt. Two moles of KOH are required to neutralize one mole of $H_2SO_4{}_{(aq)}$; only one mole is used in this case. The following equation shows how it is acidic.
$$KHSO_4{}_{(aq)} \rightarrow K^+{}_{(aq)} + H^+{}_{(aq)} + SO_4^{2-}{}_{(aq)}$$
The $H^+{}_{(aq)}$ ions cause acidity.

**3.** a) A weak base
b) $NH_3{}_{(aq)}$
c) $NH_3{}_{(aq)} + H_2O_{(l)} \rightleftharpoons H_3O^+{}_{(aq)} + NH_4^+{}_{(aq)}$



1.  $2\,HCl_{(aq)} + Ca(OH)_2{}_{(aq)} \rightarrow CaCl_2{}_{(aq)} + 2\,H_2O_{(l)}$
2.  $H_2SO_4{}_{(aq)} + 2\,KOH_{(aq)} \rightarrow K_2SO_4{}_{(aq)} + 2\,H_2O_{(l)}$
3.  $Mg(OH)_2{}_{(aq)} + H_2SO_4{}_{(aq)} \rightarrow MgSO_4{}_{(aq)} + 2\,H_2O_{(l)}$
4.  $H_3PO_4{}_{(aq)} + 3\,NaOH_{(aq)} \rightarrow Na_3PO_4{}_{(aq)} + 3\,H_2O_{(l)}$
5.  $CH_3COOH_{(aq)} + KOH_{(aq)} \rightarrow CH_3COOK_{(aq)} + H_2O_{(l)}$
6.  $2\,H_3PO_4{}_{(aq)} + 3\,Ba(OH)_2{}_{(aq)} \rightarrow Ba_3(PO_4)_2{}_{(aq)} + 6\,H_2O_{(l)}$
7.  $2\,LiOH_{(aq)} + H_2SO_4{}_{(aq)} \rightarrow Li_2SO_4{}_{(aq)} + 2\,H_2O_{(l)}$
8.  $H_2SO_3{}_{(aq)} + Ca(OH)_2{}_{(aq)} \rightarrow CaSO_3{}_{(aq)} + 2\,H_2O_{(l)}$
9.  $2\,HClO_3{}_{(aq)} + Mg(OH)_2{}_{(aq)} \rightarrow Mg(ClO_3)_2{}_{(aq)} + 2\,H_2O_{(l)}$
10. $3\,HBrO_3{}_{(aq)} + Al(OH)_3{}_{(aq)} \rightarrow Al(BrO_3)_3{}_{(aq)} + 3\,H_2O_{(l)}$

*Problem:*

**1**. What volume 0.200 mol/L of $HCl_{(aq)}$ is needed to neutralize 20.0 mL of 0.100 mol/L $Ca(OH)_{2(aq)}$?

*Given:*

Concentration of $HCl_{(aq)}$ = 0.200 mol/L

Concentration of $Ca(OH)_{2(aq)}$ = 0.100 mol/L

Volume of $Ca(OH)_{2(aq)}$ = 20.0 mL = $\dfrac{20.0\ mL}{1000\ mL/L}$ = 0.0200 L

*Required:*

The volume of 0.200 mol/L $HCl_{(aq)}$ that is required to neutralize 20.0 mL of 0.100 mol/L $Ca(OH)_{2(aq)}$

*Steps to be taken:*

**Step 1.** Balance an equation between $HCl_{(aq)}$ and $Ca(OH)_{2(aq)}$.

$$\underset{\textbf{Required}}{2\ HCl_{(aq)}} + \underset{\textbf{Given}}{Ca(OH)_{2(aq)}} \rightarrow CaCl_{2(aq)} + 2\ H_2O_{(l)}$$

**Step 2.** Since the concentration and volume of $Ca(OH)_{2(aq)}$ are given, it becomes the *given reagent* and its number of moles can be calculated.

Using the formula, $\mathbf{n = C \times V}$ we get

$$\mathbf{n} = 0.100\ mol/L \times 0.0200\ L = \mathbf{2.00 \times 10^{-3}\ mol\ Ca(OH)_{2(aq)}}$$

$$n = C \times V$$

**Step 3.** Use the number of moles of $Ca(OH)_{2(aq)}$ to calculate the number of moles of $HCl_{(aq)}$.

$$\frac{Given}{Required} = \frac{1}{2} = \frac{2.00 \times 10^{-3}\ mol\ Ca(OH)_{2(aq)}}{X\ mol\ HCl(aq)}$$

$$X = \mathbf{4.00 \times 10^{-3}\ mol\ HCl_{(aq)}}$$

**Step 4.** Since the number of moles and concentration of the $HCl_{(aq)}$ are known, its volume can be found using the following equation:

$$V = \frac{n}{C} = \frac{4.00 \times 10^{-3}\ mol}{0.200\ mol/L} = \mathbf{2.00 \times 10^{-2}\ L = 20.0\ mL}$$

$$V = \frac{n}{C}$$

*Problem:*

**2**. What volume of 0.010 mol/L $Na_2CO_{3(aq)}$ must be used to neutralize 100.0 mL of 1.00 mol/L $HCl_{(aq)}$?

*Provided:*

Concentration of $HCl_{(aq)}$ = 1.00 mol/L

Volume of $HCl_{(aq)}$ = 100.0 mL = $\dfrac{100.0\ mL}{1000\ mL/L}$ = 0.1000 L

Concentration of $Na_2CO_{3(aq)}$ = 0.010 mol/L

*Required:*

The volume of 0.010 mol/L $Na_2CO_{3(aq)}$ must be used to neutralize 100.0 mL of 1.00 mol/L $HCl_{(aq)}$.

*Steps to be taken:*

**Step 1.** Balance an equation between $HCl_{(aq)}$ and $Na_2CO_{3(aq)}$

$$\underset{\textbf{Given}}{2\ HCl_{(aq)}} + \underset{\textbf{Required}}{Na_2CO_{3(aq)}} \rightarrow 2\ NaCl_{(aq)} + H_2O_{(l)} + CO_{2(g)}$$

**Step 2.** Since the concentration and volume of $HCl_{(aq)}$ are given, it becomes the *given reagent* and its number of moles can be calculated.

Using the formula, $\mathbf{n = C \times V}$ we get

$$\mathbf{n} = 1.00 \text{ mol/L} \times 0.100 \text{ L} = 0.100 \text{ mol } HCl_{(aq)}$$

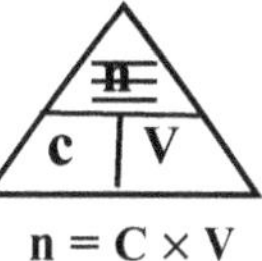


**Step 3.** Use the number of moles of $HCl_{(aq)}$ to calculate the number of moles of $Na_2CO_{3(aq)}$.

$$\frac{\text{Given}}{\text{Required}} = \frac{2}{1} = \frac{0.100 \text{ mol } HCl_{(aq)}}{X \text{ mol } Na_2CO_{3(aq)}}$$

$$2\,X = 0.100 \text{ mol } Na_2CO_{3(aq)} \ = \ X = 0.0500 \text{ mol } Na_2CO_{3(aq)}$$

**Step 4.** Since the number of moles and concentration of the $Na_2CO_{3(aq)}$ are known, its volume can be found using the following equation:

$$V = \frac{n}{C} = \frac{0.0500 \text{ mol}}{0.010 \text{ mol/L}} = 5.0 \text{ L}$$



## *Problem:*

**3.** What mass of antacid tablet, made of $Mg(OH)_{2(s)}$ is needed to neutralize 100.0 mL of stomach acid, having $HCl_{(aq)}$ of concentration 0.100 mol/L?

*Given:*

Concentration of $HCl_{(aq)} = 0.100$ mol/L

Volume of $HCl_{(aq)} = 100.0 \text{ mL} = \dfrac{100.0 \text{ mL}}{1000 \text{ mL/L}} = 0.1000 \text{ L}$

Formula for the antacid: $Mg(OH)_2$

*Required:*

The mass of antacid tablet, made of $Mg(OH)_{2(s)}$, that must be used to neutralize 100.0 mL of 2.00 mol/L $HCl_{(aq)}$ in the stomach.

*Steps to be taken:*

**Step 1.** Balance an equation between $HCl_{(aq)}$ and $Mg(OH)_{2(aq)}$.

| Given | | Required | | |
|---|---|---|---|---|
| $2 \text{ HCl}_{(aq)}$ | $+$ | $Mg(OH)_{2(s)}$ | $\rightarrow$ | $MgCl_{2(aq)} \ + \ 2 \text{ H}_2O_{(l)}$ |

**Step 2.** Since the concentration and volume of $HCl_{(aq)}$ are given, it becomes the *given reagent* and its number of moles can be calculated.

Using the formula, $\mathbf{n = C \times V}$ we get

$$\mathbf{n} = 0.100 \text{ mol/L} \times 0.100 \text{ L} = 0.0100 \text{ mol } HCl_{(aq)}$$

**Step 3.** Use the number of moles of $HCl_{(aq)}$ to calculate the number of moles of $Na_2CO_{3(aq)}$.

$$\frac{\text{Given}}{\text{Required}} = \frac{2}{1} = \frac{0.0100 \text{ mol } HCl(aq)}{X \text{ mol } Mg(OH)_{2(s)}}$$

$$2\,X = 0.0100 \text{ mol } Mg(OH)_{2(s)}$$

$$= X = 0.00500 \text{ mol } Mg(OH)_{2(s)}$$

**Step 4.** Find the mass of 0.00500 mol Mg(OH)$_{2(s)}$. To do this, the molar mass of 0.00500 mol Mg(OH)$_2$ must first be found.

$$M = 1 \times 24.31 \text{ g/mol Mg} + 2 \times 16.00 \text{ g/mol O} + 2 \times 1.01 \text{ g/mol H} = \textbf{58.33 g/mol}$$

Using the formula, **m = n × M**

$$m = 0.00500 \text{ mol} \times 58.33 \text{ g/mol} = \textbf{0.291 g Mg(OH)}_{2(s)}$$

m = n × M

## *Problem:*

4.  If acid rain has 0.0100 mol/L H$_2$SO$_{3(aq)}$ as its main ingredient, what volume of 0.050 mol/L of Na$_2$CO$_{3(aq)}$ is needed to neutralize 10.0 L of this type of rainwater?

*Provided:*
Concentration of H$_2$SO$_{3(aq)}$ = 0.0100 mol/L
Volume of H$_2$SO$_{3(aq)}$ = 10.0 L
Concentration of Na$_2$CO$_{3(aq)}$ = 0.050 mol/L

*Required:*
The volume of 0.050 mol/L of Na$_2$CO$_{3(aq)}$ is needed to neutralize 10.0 L of 0.010 mol/L H$_2$SO$_{3(aq)}$?

*Steps to be taken:*
**Step 1.** Balance an equation between and H$_2$SO$_{3(aq)}$ and Na$_2$CO$_{3(aq)}$

$$\underset{\textbf{Given}}{H_2SO_{3(aq}} + \underset{\textbf{Required}}{Na_2CO_{3(aq)}} \rightarrow Na_2SO_{3(aq} + H_2O_{(l)} + CO_{2(g)}$$

**Step 2.** Since the concentration and volume of H$_2$SO$_{3(aq)}$ are given, it becomes the *given reagent* and its number of moles can be calculated.
Using the formula, **n = C × V** we get

$$n = 0.0100 \text{ mol/L} \times 10.0 \text{ L} = \textbf{0.100 mol H}_2\textbf{SO}_{3(aq)}$$

n = C × V

**Step 3.** Use the number of moles of H$_2$SO$_{3(aq)}$ to calculate the number of moles of Na$_2$CO$_{3(aq)}$.

$$\frac{\text{Given}}{\text{Required}} = \frac{1}{1} = \frac{0.100 \text{ mol } H_2SO_{3(aq)} \text{ (aq)}}{X \text{ mol } Na_2CO_{3(aq)}}$$

$$\textbf{X = 0.100 mol Na}_2\textbf{CO}_{3(aq)}$$

**Step 4.** Since the number of moles and concentration of the Na$_2$CO$_{3(aq)}$ are known, its volume can be found using the following equation:

$$V = \frac{n}{C} = \frac{0.100 \text{ mol}}{0.050 \text{ mol/L}} = \textbf{2.0 L Na}_2\textbf{CO}_{3(aq)}$$

$$V = \frac{n}{C}$$

## *Problem:*

5. Acetic acid, CH$_3$COOH$_{(aq)}$ is sold under name vinegar.  If 10.0 mL of 0.100 mol/L NaOH$_{(aq)}$ is required to neutralize 8.00 mL of vinegar, what is the concentration of the vinegar?

*Given:*
Concentration of NaOH$_{(aq)}$ = 0.100 mol/L
Volume of NaOH$_{(aq)}$ = 10.0 mL = $\dfrac{10.0 \text{ mL}}{1000\text{mL/L}}$ = 0.0100 L
Volume of CH$_3$COOH$_{(aq)}$ = 8.0 mL = $\dfrac{8.00 \text{ mL}}{1000 \text{ mL/L}}$ = 8.00 x 10$^{-3}$ L

*Required:*
The concentration of $CH_3COOH_{(aq)}$

*Steps to be taken:*
**Step 1.** Balance an equation between the following two reactants:

$$\underset{\text{Required}}{NaOH_{(aq)}} + \underset{\text{Given}}{CH_3COOH_{(aq)}} \rightarrow CH_3COONa_{(aq)} + H_2O_{(l)}$$

**Step 2.** Since the concentration and volume of $NaOH_{(aq)}$ are given, it becomes the *given reagent,* and its number of moles can be calculated.

Using the formula, $\mathbf{n = C \times V}$ we get

$$\mathbf{n = 0.100 \ mol/L \times 0.0100 \ L = 1.00 \times 10^{-3} \ mol \ NaOH_{(aq)}}$$

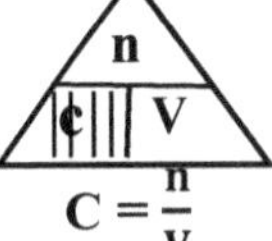


**Step 3.** Use the number of moles of $NaOH_{(aq)}$ to calculate the number of moles of $CH_3COOH_{(aq)}$

$$\frac{\text{Given}}{\text{Required}} = \frac{1}{1} = \frac{1.00 \times 10^{-3} \ mol \ NaOH_{(aq)}}{X \ mol \ CH_3COOH(aq)}$$

$$\mathbf{X = 1.00 \times 10^{-3} \ mol \ CH_3COOH_{(aq)}}$$

**Step 4.** Since the concentration and concentration of the $CH_3COOH_{(aq}$ are known, its volume can be found using the following equation:

$$\mathbf{C = \frac{n}{V}} = \frac{1.00 \times 10^{-3} \ mol}{8.00 \times 10^{-3} \ L} = \mathbf{0.130 \ mol/L \ CH_3COOH_{(aq)}}$$



## Problem:

**6.** In an experiment, 10.0 mL of 0.100 mol/L $NaOH_{(aq)}$ was needed to neutralize 10.0 mL of a $H_2SO_{4(aq)}$ solution. Calculate the concentration of the acid $H_2SO_{4(aq)}$ solution.

*Provided:*
Concentration of $NaOH_{(aq)}$ = 0.100 mol/L

Volume of $NaOH_{(aq)}$ = 10.0 mL = $\dfrac{10.0 \ mL}{1000 \ mL/L}$ = 0.0100 L

Volume of $H_2SO_{4(aq)}$ = 10.0 mL = $\dfrac{10.0 \ mL}{1000.0 \ mL/L}$ = 0.0100 L

*Required:*
The concentration of = $H_2SO_{4(aq)}$

*Steps to be taken:*
**Step 1.** Balance an equation between the two reactants.

$$\underset{\text{Given}}{2 \ NaOH_{(aq)}} + \underset{\text{Required}}{H_2SO_{4(aq)}} \rightarrow Na_2SO_{4(aq)} + 2 \ H_2O_{(l)}$$

**Step 2.** Since the concentration and volume of $NaOH_{(aq)}$ are given, it becomes the *given reagent* and its number of moles can be calculated.

Using the formula, $\mathbf{n = C \times V}$ we get

$$\mathbf{n = 0.100 \ mol/L \times 0.0100 \ L = 1.00 \times 10^{-3} \ mol \ NaOH_{(aq)}}$$



**Step 3.** Use the number of moles of $NaOH_{(aq)}$ to calculate the number of moles of $H_2SO_{4(aq)}$.

$$\frac{\text{Given}}{\text{Required}} = \frac{2}{1} = \frac{1.00 \times 10^{-3} \ mol \ NaOH_{(aq)}}{X \ mol \ H_2SO_4 \ (aq)}$$

$$\mathbf{2 \ X = 1.00 \times 10^{-3} \ mol \ H_2SO_{4(aq)} \ = X = 5.00 \times 10^{-4} \ mol \ H_2SO_{4(aq)}}$$

**Step 4.** Since the number of moles and volume of the $H_2SO_{4(aq)}$ are known, its concentration can be found using the following equation:

$$C = \frac{n}{V} = \frac{5.00 \times 10^{-4}\ \text{mol}}{0.0100\ \text{L}} = 0.0500\ \text{mol/L}\ H_2SO_{4(aq)}$$

## *Problem:*

7. A student mixed 100.0 mL of 0.200 mol/L $H_2SO_{4(aq)}$ with 400.0 mL of 0.200 mol/L $NaOH_{(aq)}$ and the solution produced was found to basis. What volume of 0.100 mol/L $HCl_{(aq)}$ is required to neutralize the excess base?

## *Provided:*

Concentration of $NaOH_{(aq)}$ = 0.20 mol/L

Volume of $NaOH_{(aq)}$ = 400.0 mL = $\dfrac{400.0\ \text{mL}}{1000\ \text{mL/L}}$ = 0.4000 L

Volume of $H_2SO_{4(aq)}$ = 100.0 mL = $\dfrac{100.0\ \text{mL}}{1000\ \text{mL/L}}$ = 0.1000 L

Concentration of $H_2SO_{4(aq)}$ = 0.200 mol/

## *Required:*

The volume of 0.100 mol/L $HCl_{(aq)}$ that is required to neutralize the excess base?

## *Steps to be taken:*

In this problem, it is told that the $NaOH_{(aq)}$ is *the excess reagent*, so the $H_2SO_{4(aq)}$ is *the limiting reagent* for the first reaction. To solve this problem, find the number of moles of both reagents and then use the number of moles of $H_2SO_{4(aq)}$ to find how much of the $NaOH_{(aq)}$ was neutralized by it. This will enable us to know how much of $NaOH_{(aq)}$ was in excess. Knowing this, the amount and volume of 0.100 mol/L $HCl_{(aq)}$ required to neutralize it can be calculated.

**Step 1.** Balance an equation between $NaOH_{(aq)}$ and $H_2SO_{4(aq)}$

$$2\ NaOH_{(aq)} \ + \ H_2SO_{4(aq)} \ \rightarrow \ Na_2SO_{4(aq)} \ + \ 2\ H_2O_{(l)}$$

**Step 2.** Find the number of moles of **$NaOH_{(aq)}$ used**.

Using the formula, **n = C × V** we get

$$\mathbf{n} = 0.200\ \text{mol/L} \times 0.4000\ \text{L} = \mathbf{0.0800\ mol\ NaOH_{(aq)}\ (Used)}$$

**Step 3.** Find the number of moles of **$H_2SO_{4(aq)}$ used**.

Using the formula, **n = C × V** we get

$$\mathbf{n} = 0.200\ \text{mol/L} \times 0.1000\ \text{L} = \mathbf{0.0200\ mol\ H_2SO_{4(aq)}\ (Used)}$$

**Step 4.** Since the $H_2SO_{4(aq)}$ is the limiting reagent, it is called *the given reagent*. It is now used to calculate how $NaOH_{(aq)}$ was used up to neutralize it.

$$\frac{\text{Given}}{\text{Required}} = \frac{1}{2} = \frac{0.0200\ \text{mol}\ H_2SO_4\ \text{(aq)}}{X\ \text{mol NaOH (aq)}}$$

$$X = \mathbf{0.0400\ mol\ NaOH_{(aq)}}$$

**Step 5.** Find the amount of **$NaOH_{(aq)}$** that is in **excess**.

$$= 0.0800\ \text{mol}\ (\textbf{added}) - 0.0400\ \text{mol}\ (\textbf{neutralized}) = \mathbf{0.0400\ mol\ NaOH_{(aq)}}$$

**Step 6.** Balance an equation to show the reaction between $NaOH_{(aq)}$ and $HCl_{(aq)}$. In this case, the $NaOH_{(aq)}$ becomes *the given reagent* and the $HCl_{(aq)}$, *the required one.*

$$\underset{\text{Given}}{NaOH_{(aq)}} \quad + \quad \underset{\text{Required}}{HCl_{(aq)}} \quad \rightarrow \quad NaCl_{(aq)} \quad + \quad H_2O_{(l)}$$

**Step 7.** Find the number of moles of $HCl_{(aq)}$ required to neutralize **0.0400 mol of excess $NaOH_{(aq)}$.**

$$\frac{\text{Given}}{\text{Required}} = \frac{1}{1} = \frac{0.0400 \text{ mol NaOH}_{(aq)}}{X \text{ mol HCl(aq)}}$$

$$X = \textbf{0.0400 mol HCl}_{(aq)}$$

**Step 8.** Since the number of moles and concentration of $HCl_{(aq)}$ are known, its volume can be found using the following equation:

$$V = \frac{n}{C} = \frac{0.0400 \text{ mol}}{0.100 \text{ mol/L}} = \textbf{0.400 L} = \textbf{4.00} \times \textbf{10}^2 \textbf{ mL}$$

## *Problem:*

**8.** What mass of $Mg_{(s)}$ would be required to neutralize the excess acid from a mixture of 200.0 mL of 0.100 mol/L $NaOH_{(aq)}$ and 250.0 mL of 0.100 mol/L $H_2SO_{4(aq)}$? Also, find the mass of hydrogen gas that would be produced.

## *Provided:*

$$\text{Volume of NaOH}_{(aq)} = 200.0 \text{ mL} = \frac{200.0 \text{ mL}}{1000 \text{ mL/L}} = 0.2000 \text{ L}$$

Concentration of $NaOH_{(aq)}$ = 0.100 mol/L

$$\text{Volume of H}_2SO_{4(aq)} = 250.0 \text{ mL} = \frac{250.0 \text{ mL}}{1000 \text{ mL/L}} = 0.2500 \text{ L}$$

Concentration of $H_2SO_{4(aq)}$ = 0.100 mol/L

## *Required:*

*The mass of $Mg_{(s)}$ would be required to neutralize the excess acid from a mixture of 200.0 mL of 0.100 mol/L $NaOH_{(aq)}$ and 250.0 mL of 0.100 mol/L $H_2SO_{4(aq)}$.

*The mass of $H_{2(g)}$ produced from the reaction between the excess $H_2SO_{4(aq)}$ and $Mg_{(s)}$ used.

## *Steps to be taken:*

In this problem, it is told that the **$H_2SO_{4(aq)}$** is *the excess reagent*, so the **$NaOH_{(aq)}$** is *the limiting reagent.* To solve this problem, find the number of moles of both reagents and then use the number of moles of $NaOH_{(aq)}$ to find how much of the $H_2SO_{4(aq)}$ was neutralized by it. This will enable us to know how much of $H_2SO_{4(aq)}$ was in excess, and knowing this, the mass of $Mg_{(s)}$ required to neutralize it can be calculated.

**Step 1.** Balance an equation between $NaOH_{(aq)}$ and $H_2SO_{4(aq)}$

$$\underset{\text{Given}}{2 \text{ NaOH}_{(aq)}} \quad + \quad \underset{\text{Required}}{H_2SO_{4(aq)}} \quad \rightarrow \quad Na_2SO_{4(aq)} \quad + \quad 2 \text{ H}_2O_{(l)}$$

**Step 2.** Find the number of moles of $NaOH_{(aq)}$ used.

Using the formula, $n = C \times V$ we get

$$n = 0.100 \text{ mol/L} \times 0.200 \text{ L} = \textbf{0.0200 mol NaOH}_{(aq)} \textbf{ (Used)}$$

**Step 3.** Find the number of moles of $H_2SO_{4(aq)}$ used.

Using the formula, $\mathbf{n = C \times V}$ we get

$\mathbf{n = 0.100 \ mol/L \times 0.250 \ L = 0.0250 \ mol \ H_2SO_{4(aq)} \ (Used)}$

**Step 4.** Since the $NaOH_{(aq)}$ is the limiting reagent, and it is called *the given reagent*. It is now used to calculate how $H_2SO_{4(aq)}$ was used up to neutralize it.

$$\frac{\text{Given}}{\text{Required}} = \frac{2}{1} = \frac{0.0200 \ mol \ NaOH \ (aq)}{X \ mol \ H_2SO_4 \ (aq)}$$

$\mathbf{2 \ X = 0.0200 \ mol \ H_2SO_{4(aq)} = X = 0.0100 \ mol \ H_2SO_{4(aq)} \ (used)}$

**Step 5.** Find the amount of that $\mathbf{H_2SO_{4(aq)}}$ is in **excess**.

$= 0.0250 \ mol \ (\mathbf{added}) - 0.0100 \ mol \ (\mathbf{used}) = \mathbf{0.0150 \ mol \ H_2SO_{4(aq)}}$

**Step 6.** Balance an equation to show the reaction between $Mg_{(s)}$ and $H_2SO_{4(aq)}$. In this case, the $H_2SO_{4(aq)}$ becomes *the given reagent* and the $Mg_{(s)}$, *the required one*.

$$\overset{\textbf{Given}}{H_2SO_{4(aq)}} \quad + \quad \overset{\textbf{Required}}{Mg_{(s)}} \ \longrightarrow \quad MgSO_{4(aq)} \ + \ H_{2(g)}$$

**Step 7.** Find the number of moles of $Mg_{(s)}$ that is required to neutralize **0.0150 mol of excess** $\mathbf{H_2SO_{4(aq)}}$.

$$\frac{\text{Given}}{\text{Required}} = \frac{1}{1} = \frac{0.0150 \ mol \ H_2SO_4 \ (aq)}{X \ mol \ Mg_{(s)}}$$

$$X = \mathbf{0.0150 \ mol \ Mg_{(s)}}$$

**Step 8.** Find the mass of **0.0150 moles of** $\mathbf{Mg_{(s)}}$.

Using the formula, $\mathbf{m = n \times M}$ we get

Mass $= 0.0150 \ mol \times 24.31 \ g/mol \ Mg = \mathbf{0.364 \ g \ Mg_{(s)}}$

**Mass of $\mathbf{H_{2(g)}}$ produced**

**Step 9.** Use the balance equation between $H_2SO_{4(aq)}$ and $Mg_{(s)}$ to find the number of moles of $H_{2(g)}$ produced. In this case, the number of moles of $Mg_{(s)}$ is used and it thus becomes *the given reagent* and the $H_{2(g)}$ becomes *the required reagent*.

$$H_2SO_{4(aq)} \ + \ \overset{\textbf{Given}}{Mg_{(s)}} \ \longrightarrow \quad Mg \ SO_{4(aq)} \ + \ \overset{\textbf{Required}}{H_{2(g)}}$$

$$\frac{\text{Given}}{\text{Required}} = \frac{1}{1} = \frac{0.0150 \ mol \ Mg_{(s)}}{X \ mol \ H_{2(g)}}$$

$$X = \mathbf{0.0150 \ mol \ H_{2(g)}}$$

**Step 10.** Find the mass of $0.0150$ moles of $H_{2(g)}$.

Using the formula, $\mathbf{m = n \times M}$ we get

Mass $= 0.0150 \ mol \times 2.02 \ g/mol \ H_{2(g)} = \mathbf{0.0303 \ g \ H_{2(g)}}$

*Problem:*

**9.** In an effort to neutralize an acid spill of 500.0 mL of 0.100 mol/L $HNO_{3(aq)}$, a student poured 200.0 mL of 0.200 mol/L $KOH_{(aq)}$ onto it, but subsequently found that the resulting mixture was still  acidic. What mass of $NaHCO_{3(s)}$ is required to react with any excess acid from this reaction.

*Provided:*
Concentration of $HNO_{3(aq)}$ = 0.100 mol/L

Volume of $HNO_{3(aq)}$ = 500.0 mL = $\dfrac{500.0 \text{ mL}}{1000 \text{ mL/L}}$ = 0.5000 L

Volume of $KOH_{(aq)}$ = 200.0 mL = $\dfrac{200.0 \text{ mL}}{1000 \text{ mL/L}}$ = 0.2000 L

Concentration of $KOH_{(aq)}$ = 0.200 mol/L

*Required:*
The mass of $NaHCO_{3(s)}$ is required to react with any excess acid?

In this problem, it is told that the $HNO_{3(aq)}$ is *the excess reagent,* so the $KOH_{(aq)}$ is *the limiting reagent,* for the first reaction. To solve this problem, find the number of moles of both reagents and then use the number of moles of $KOH_{(aq)}$ to find how much of the $HNO_{3(aq)}$ was neutralized by it. This will enable us to know how much of $HNO_{3(aq)}$ was in excess, and knowing this, the mass of $NaHCO_{3(s)}$ required to neutralize it can be calculated.

*Steps to be taken:*
**Step 1.** Balance an equation between $KOH_{(aq)}$ and $HNO_{3(aq}$

| Given | | Required | | | |
|---|---|---|---|---|---|
| $KOH_{(aq)}$ | + | $HNO_{3(aq)}$ | $\rightarrow$ | $HNO_{3(aq)}$ | + $H_2O_{(l)}$ |

**Step 2.** Find the number of moles of $KOH_{(aq)}$.

Using the formula, **n = C × V** we get

**n** = 0.200 mol/L × 0.200 L = **0.0400 mol $KOH_{(aq)}$ (Used)**

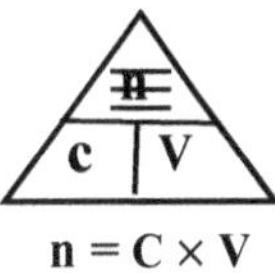
$n = C \times V$

**Step 3.** Find the number of moles of $HNO_{3(aq)}$.
Using the formula, **n = C × V** we get

**n** = 0.100 mol/L × 0.500 L = **0.0500 mol $HNO_{3(aq)}$ (Used)**

**Step 4.** Use the number of moles of $KOH_{(aq)}$ to calculate the number of moles of $HNO_{3(aq)}$ that are required to neutralize it; $KOH_{(aq}$ becomes *the given reagent* and $HNO_{3(aq)}$ *the required one.*

$$\frac{\text{Given}}{\text{Required}} = \frac{1}{1} = \frac{0.0400 \text{ mol } KOH_{(aq)}}{X \text{ mol } HNO_3 \text{ (aq)}}$$
$$X = 0.0400 \text{ mol } HNO_{3(aq)}$$

**Step 5.** Find the number of moles of **excess $HNO_{3(aq)}$**

= 0.0500 mol (**added**) - 0.0400 mol (**neutralized**) = **0.0100 mol $HNO_{3(aq)}$**

**Step 6.** Balance an equation to show the reaction between and $HNO_{3(aq)}$ $NaHCO_{3(s)}$. In this case, the excess 0.0100 mol $HNO_{3(aq)}$ becomes *the given reagent* and the $NaHCO_{3(s)}$, *the required one.*

| Given | | Required | | | | |
|---|---|---|---|---|---|---|
| $HNO_{3(aq)}$ | + | $NaHCO_{3(s)}$ | $\rightarrow$ | $NaNO_{3(aq)}$ + | $H_2O_{(l)}$ + | $CO_{2(g)}$ |

**Step 7.** Find the number of moles of $NaHCO_{3(s)}$ required to react with 0.01 mol $HNO_{3(aq)}$.

$$\frac{\text{Given}}{\text{Required}} = \frac{1}{1} = \frac{0.0100 \text{ mol } HNO_{3(aq)}}{X \text{ mol } NaHCO_{3(s)}}$$
$$X = 0.0100 \text{ mol } NaHCO_{3(s)}$$

**Step 8.** Find the mass of 0.01 mol $NaHCO_{3(s)}$. To do this first find the molar mass of 84.01 g/mol

$M = 1 \times 22.99$ g/mol Na $+ 1 \times 1.01$ g/mol H $+ 1 \times 12.01$ g/mol C $+ 3 \times 16.00$ g/mol O $= \mathbf{\textit{84.01 g/mol}}$

Using the the fomula, $\mathbf{m = n \times M}$ we get

Mass $= 0.0100$ mol $\times 84.01$ g/mol $NaHCO_{3(s)}$

$= \mathbf{0.840\ g\ NaHCO_{3(s)}}$

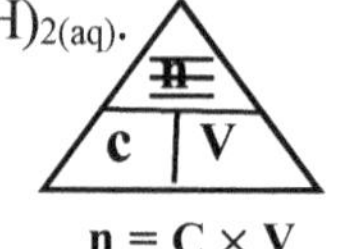

$\mathbf{m = n \times M}$

## *Problem:*

**9.** 100.0 mL of 0.100 mol/L $HCl_{(aq)}$ is mixed with 140.0 mL of a solution of $Ba(OH)_{2(aq)}$. A pH meter placed in the mixture shows that the mixture is basic and that it required a further 400 mL of 0.0500 mol/L $HCl_{(aq)}$ for neutralization. Calculate the concentration of the $Ba(OH)_{2(aq)}$ solution.

*Provided:*

**Initial addition**

Concentration of $HCl_{(aq)} = 0.100$ mol/L

Volume of $HCl_{(aq)} = 100.0$ mL $= \dfrac{100.0 \text{ mL}}{1000 \text{ mL/L}} = 0.1000$ L

Volume of $Ba(OH)_{2(aq)} = 140.0$ mL $= \dfrac{140.0 \text{ mL}}{1000 \text{ mL/L}} = 0.1400$ L

**Second addition**

Concentration of $HCl_{(aq)} = 0.050$ mol/L

Volume of $HCl_{(aq)} = 40.0$ mL $= \dfrac{40.0 \text{ mL}}{1000 \text{ mL/L}} = 0.0400$ L

*Required:* The concentration of the $Ba(OH)_{2(aq)}$.

In this problem, it is given that the amount of the $Ba(OH)_{2(aq)}$ was *the excess reagent* for the first addition of $HCl_{(aq)}$. Knowing the volume and concentration of the $HCl_{(aq)}$ used in the first addition, the amount of $Ba(OH)_{2(aq)}$ can be calculated. Also, knowing the volume and concentration of the $HCl_{(aq)}$ used in the second addition, the excess amount of $Ba(OH)_{2(aq)}$ can therefore be calculated. Thus the total number of moles of $Ba(OH)_{2(aq)}$ present in the solution is found from the sum of these two. Knowing the total number of moles $Ba(OH)_2$, its concentration can be calculated since its volume is provided.

*Steps to be taken:*

**Step 1.** Balance an equation between $HCl_{(aq)}$ and $Ba(OH)_{2(aq)}$.

| Given | | Required | | | |
| --- | --- | --- | --- | --- | --- |

$2\ HCl_{(aq)} \quad + \quad Ba(OH)_{2(aq)} \quad \rightarrow \quad BaCl_{2(aq)} \quad + \quad 2\ H_2O_{(l)}$

**Step 2.** Find the number of moles of $HCl_{(aq)}$ used to neutralize the excess $Ba(OH)_{2(aq)}$.

Using the formula, $\mathbf{n = C \times V}$ we get

$\mathbf{n = 0.050}$ mol/L $\times 0.0400$ L $= \mathbf{2.00 \times 10^{-3}}$ **mol** $\mathbf{HCl_{(aq)}}$

$\mathbf{n = C \times V}$

**Step 3.** Use the number of moles of $HCl_{(aq)}$ to calculate the number of moles of $Ba(OH)_{2(aq)}$ in excess.

$$\dfrac{\text{Given}}{\text{Required}} = \dfrac{2}{1} = \dfrac{2.00 \times 10^{-3} \text{ mol } HCl_{(aq)}}{\text{X mol } Ba(OH)_{2(aq)}}$$

$2\ X = \mathbf{2.00 \times 10^{-3}}$ **mol** $\mathbf{Ba(OH)_{2(aq)}} = X = \mathbf{1.0 \times 10^{-3}}$ **mol** $\mathbf{Ba(OH)_{2(aq}}$

**Step 4.** Find the number of moles of moles of $HCl_{(aq)}$ initially added.

Using the formula, $\mathbf{n = C \times V}$ we get

$\mathbf{n = 0.100}$ mol/L $\times 0.100$ L $= \mathbf{0.0100}$ **mol** $\mathbf{HCl_{(aq)}}$

**Step 5.** Use the number of moles of $HCl_{(aq)}$ to calculate the number of moles of $Ba(OH)_{2(aq)}$ that was initially neutralized.

$$\frac{Given}{Required} = \frac{2}{1} = \frac{0.0100 \text{ mol } HCl_{(aq)}}{X \text{ mol } Ba(OH)_{2(aq)}}$$

$$2\,X = 0.0100 \text{ mol } Ba(OH)_{2(aq)} = X = 5.00 \times 10^{-3} \text{ mol } Ba(OH)_{2(aq)}$$

**Step 6.** Find the total number of moles of $Ba(OH)_{2(aq)}$ in the original volume of 140.0 mL.

$$1.0 \times 10^{-3} \text{ mol} + 5.00 \times 10^{-3} \text{ mol} = 6.00 \times 10^{-3} \text{ mol } Ba(OH)_{2(aq}$$

**Step 7.** Find the concentration of the $Ba(OH)_{2(aq)}$ of the initial solution.

Using the formula, $C = \dfrac{n}{V}$ we get

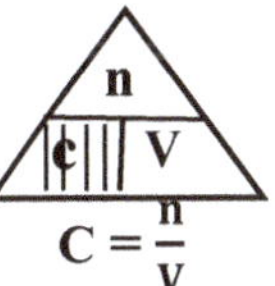

$$C = \frac{6.00 \times 10^{-3} \text{ mol}}{0.1400 \text{ L}} = 4.28 \times 10^{-2} \text{ mol/L } Ba(OH)_{2(aq)}$$

## *Problem:*

**11.** Figure A below shows how a solution of $HCl_{(aq)}$ can be reacted with a mass of powdered $CaCO_{3(s)}$ and any gas produced be collected in a balloon. This done by first placing the $CaCO_{3(s)}$ in a deflated balloon which is then secured over the mouth of an Erlenmeyer flask. The $CaCO_{3(s)}$ is the poured into the acid.

*Provided:*

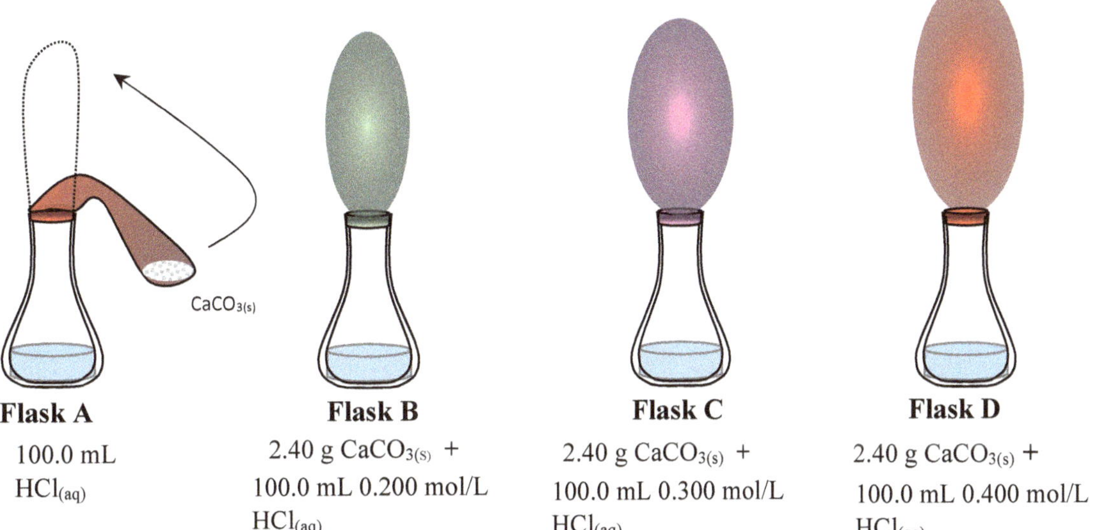


**Flask A**
100.0 mL
$HCl_{(aq)}$

**Flask B**
2.40 g $CaCO_{3(S)}$ +
100.0 mL 0.200 mol/L
$HCl_{(aq)}$

**Flask C**
2.40 g $CaCO_{3(s)}$ +
100.0 mL 0.300 mol/L
$HCl_{(aq)}$

**Flask D**
2.40 g $CaCO_{3(s)}$ +
100.0 mL 0.400 mol/L
$HCl_{(aq)}$

*Required:*

a) Write a balanced equation for the reaction between $CaCO_{3(s)}$ and $HCl_{(aq)}$.

b) What mass of $CO_{2(g)}$ is collected in each balloon?

c) In which flask is the acid completely neutralized?

d) How do you account for the differences in the volumes of the balloons?

*Strategy:*

For each of these flasks, the limiting reagent will have to be found, and then use it to find the mass of $CO_{2(g)}$ collected in each balloon. However, it can be observed that as the amount acid (**an increase in concentration**) in the flasks increased the amount of $CO_{2(g)}$ collected in each balloon also increased. It can thus be concluded that the $HCl_{(aq)}$ was the limiting reagent and the $CaCO_{3(s)}$ was the excess reagent, at least in flasks B and C. In these, the $HCl_{(aq)}$ thus becomes *the given reagent* and $CO_{2(g)}$ becomes *the required reagent*. In flasks B and C, it can be concluded for sure that the acid was completely neutralized, since an increase in the amount of it in flask D yielded more $CO_{2(g)}$; meaning that there was still an excess of $CaCO_{3(s)}$ in flask C. The same conclusion cannot be made for flask D; calculations will have to be made to determine the limiting reagent and to find out if the acid was completely neutralized or not.

*Steps to be taken:*

**Step 1.** Balance an equation between $HCl_{(aq)}$ and $CaCO_{3(s)}$ in flask B.

Given            Required
$$2\ HCl_{(aq)}\ +\ CaCO_{3(s)}\ \rightarrow\ CaCl_{2(aq)}\ +\ H_2O_{(l)}\ +\ CO_{2(g)}$$

**Step 2.** Find the number of moles of $HCl_{(aq)}$ used in flask **B**.

Using the formula, $\mathbf{n = C \times V}$ we get

$$n = 0.200\ mol/L \times 0.1000\ L = \mathbf{0.0200\ mol\ HCl_{(aq)}}$$

$$n = C \times V$$

**Step 3.** Use the number of moles of $HCl_{(aq)}$ to calculate the number of moles of $CO_{2(g)}$

$$\frac{Given}{Required} = \frac{2}{1} = \frac{0.0200\ mol\ HCl_{(aq)}}{X\ mol\ CO_{2(g)}}$$

$$\mathbf{2\ X = 0.0200\ mol\ CO_{2(g)} = X = 0.0100\ mol\ CO_{2(g)}}$$

**Step 4.** Find the mass of $CO_{2(g)}$ produced. First find the molar mass of $CO_{2(g)}$.

$$M = 1 \times 12.01\ g/mol\ C\ + 2 \times 16.00\ g/mol\ Oxygen\ = \mathbf{\textit{44.01 g/mol}}$$
Using the formula, $\mathbf{m = n \times M}$ we get

$$Mass = 0.0100\ mol \times 44.01\ g/mol = \mathbf{0.440\ g\ CO_{2(g)}}$$

$$m = n \times M$$

**Step 5.** Find the number of moles of $HCl_{(aq)}$ used in flask **C**.

Using the formula, $\mathbf{n = C \times V}$ we get

$$n = 0.300\ mol/L \times 0.100\ L = \mathbf{0.0300\ mol\ HCl_{(aq)}}$$

**Step 6.** Use the number of moles of $HCl_{(aq)}$ to calculate the number of moles of $CO_{2(g)}$.

$$\frac{Given}{Required} = \frac{2}{1} = \frac{0.0300\ mol\ HCl_{(aq)}}{X\ mol\ CO_{2(g)}}$$

$$\mathbf{2\ X = 0.0300\ mol\ CO_{2(g)} = X = 1.50 \times 10^{-2}\ mol\ CO_{2(g)}}$$

**Step 7.** Find the mass of $CO_{2(g)}$ produced.

$$Mass = 1.50 \times 10^{-2}\ mol \times 44.01\ g/mol = \mathbf{0.66\ g\ CO_{2(g)}}$$

For flask D, the limiting reagent must be determined. In this case we will choose $HCl_{(aq)}$ to determine how much $CaCO_{3(s)}$ is required to neutralize it.

**Step 8.** Balance an equation between $HCl_{(aq)}$ and $CaCO_{3(s)}$ in flask D

Given       Required
$$2\ HCl_{(aq)}\ +\ CaCO_{3(s)}\ \rightarrow\ CaCl_{2(aq)}\ +\ H_2O_{(l)}\ +\ CO_{2(g)}$$

**Step 9.** Find the number of moles of $HCl_{(aq)}$ used in flask **D**.

Using the formula, $\mathbf{n = C \times V}$ we get

$$n = 0.400\ mol/L \times 0.100\ L = \mathbf{0.0400\ mol\ HCl_{(aq)}}$$

**Step 10.** Use the number of moles of $HCl_{(aq)}$ to calculate the number of moles of $CaCO_{3(s)}$ required to neutralize it.

$$\frac{Given}{Required} = \frac{2}{1} = \frac{0.0400\ mol\ HCl_{(aq)}}{X\ mol\ CaCO_{3(s)}}$$

$$\mathbf{2X = 0.0400\ mol\ CaCO_{3(s)}\ =\ X = 0.0200\ mol\ CaCO_{3(s)}\ (Required)}$$

**Step 11.** Find the number of moles of $CaCO_{3(s)}$ in 2.40 g $CaCO_{3(s)}$ used in flask D. First find its molar mass.

$$M = 1 \times 40.08 \text{ g/mol Ca} + 1 \times 12.01 \text{ g/mol C} + 3 \times 16.00 \text{ g/mol O} = \mathbf{\textit{100.1 g/mol}}$$

Using the formula, $\mathbf{n = \dfrac{m}{M}}$ we get

$$n = \frac{2.40 \text{ g}}{100.1 \text{ g/mol}} = \mathbf{0.0240 \text{ mol } CaCO_{3(s)} \textbf{ (Used)}}$$

$$m = n \times M$$

It can be seen that the amount of $CaCO_{3(s)}$ used **(0.0240 mol)** was more than was required **(0.0200 mol)** to neutralize 0.0400 mol of $HCl_{(aq)}$; it thus the excess reagent and $HCl_{(aq)}$ is the limiting reagent.

**Step 12.** Use the number of moles of $HCl_{(aq)}$ to calculate the number of moles of $CO_{2(g)}$ produced in flask D.

| Given | | | | Required |
|---|---|---|---|---|

$$2 \text{ HCl}_{(aq)} + \text{CaCO}_{3(s)} \rightarrow \text{CaCl}_{2(aq)} + \text{H}_2\text{O}_{(l)} + \text{CO}_{2(g)}$$

$$\frac{\text{Given}}{\text{Required}} = \frac{2}{1} = \frac{0.0400 \text{ mol HCl}_{(aq)}}{\text{X mol CO}_{2(g)}}$$

$$\mathbf{2 \text{ X} = 0.0400 \text{ mol } CO_{2(g)} = \text{X} = 0.0200 \text{ mol } CO_{2(g)}}$$

**Step 13.** Find the mass of $CO_{2(g)}$ produced.

$$\text{Mass} = 0.0200 \text{ mol} \times 44.01 \text{ g/mol } CO_{2(g)} = \mathbf{0.880 \text{ g } CO_{2(g)}}$$

c) Since the mass of $CaCO_{3(s)}$ was in excess in flask D, which had the largest amount of $HCl_{(aq)}$ and complete neutralization occurred there; the same must happen in the other flasks which had smaller amounts of $HCl_{(aq)}$.

d) There was a progressive increase in the volume of $CO_{2(g)}$ from flasks B to D because as more $HCl_{(aq)}$ was added, it was able to react with more of the $CaCO_{3(s)}$ present to produce more $CO_{2(g)}$. This happened because in every flask, the $CaCO_{3(s)}$ was always in excess to react with the increasing amount of $HCl_{(aq)}$ added.

## Exercise 17.6 (p.287-288)

### *Problem:*

**a)** The table next shows the results obtained when a 0.100 mol/L $LiOH_{(aq)}$ is titrated against 10.0 mL aliquots of $H_2SO_{4(aq)}$ of unknown concentration:

**Table 16.11.** Volume of 0.100 mol/L $LiOH_{(aq)}$ used.

| Trial | 1 | 2 | 3 | 4 |
|---|---|---|---|---|
| Final burette reading (mL) | 10.2 | 20.3 | 30.4 | 30.5 |
| Initial burette reading (mL) | 0 | 10.2 | 20.3 | 30.4 |
| Volume of $LiOH_{(aq)}$ used (mL) | 10.2 | 10.1 | 10.1 | 10.1 |

*Provided:*

Volume of $LiOH_{(aq)}$ (to be calculated)

Concentration of $LiOH_{(aq)} = 0.100$ mol/L

$$\text{Volume of } H_2SO_{4(aq)} = 10.0 \text{ mL} = \frac{10.0 \text{ mL}}{1000.0 \text{ mL/L}} = 0.0100 \text{ L}$$

*Required:*

The concentration of the $H_2SO_{4(aq)}$.

*Strategy:*

To solve this problem, first find the average volume of the $LiOH_{(aq)}$; this will allow number of moles of $LiOH_{(aq)}$ used in the titration to be calculated. Next, balance an equation to show how $LiOH_{(aq)}$ and $H_2SO_{4(aq)}$ react; this will allow number of moles of used $H_2SO_{4(aq)}$ to be calculated. Knowing the number of moles and concentration of the $H_2SO_{4(aq)}$, its volume can be calculated.

*Steps to taken:*

**Step 1**. Find the average volume of the $LiOH_{(aq)}$. (Trial1 is discarded because of its inconsistency).

$$V = \frac{10.1 \text{ mL} + 10.1 \text{ mL} + 10.1 \text{ mL}}{3} = 10.1 \text{ mL} = \frac{10.1 \text{ mL}}{1000 \text{ mL/L}} = 0.0101 \text{ L}$$

**Step 2**. Find the number of moles of $LiOH_{(aq)}$ used.

Using the formula, $\mathbf{n = C \times V}$ we get

$$\mathbf{n = 0.100 \text{ mol/L} \times 0.0101 \text{ L} = 1.01 \times 10^{-3} \text{ mol } LiOH_{(aq)}}$$

$n = C \times V$

**Step 3.** Balance an equation to show how $LiOH_{(aq)}$ and $H_2SO_{4(aq)}$ react.

**Given**        **Required**

$2 \text{ } LiOH_{(aq)} \quad + \quad H_2SO_{4(aq)} \quad \rightarrow \quad Li_2SO_{4(aq)} \quad + \quad 2 \text{ } H_2O_{(l)}$

**Step 4.** Find the number of moles of $H_2SO_{4(aq)}$ used.

$$\frac{\text{Given}}{\text{Required}} = \frac{2}{1} = \frac{1.01 \times 10^{-3} \text{ mol } LiOH_{(aq)}}{X \text{ mol } H_2SO_{4(aq)}}$$

$$\mathbf{2\,X = 1.01 \times 10^{-3} \text{ mol } H_2SO_{4(aq)} = X = 5.05 \times 10^{-4} \text{ mol } H_2SO_{4(aq)}}$$

**Step 5.** Find the concentration of the $H_2SO_{4(aq)}$.

Using the formula, $\mathbf{c = \dfrac{n}{V}}$ we get

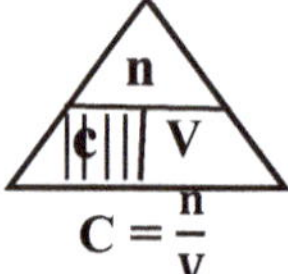


$C = \dfrac{n}{V}$

$$C = \frac{5.05 \times 10^{-4} \text{ mol}}{1.00 \times 10^{-2} \text{ L}} = \mathbf{5.05 \times 10^{-2} \text{ mol/L } H_2SO_{4(aq)}}$$

## Problem:

**b)** The table below shows the results obtained when a 0.200 mol/L $Ba(OH)_{2(aq)}$ is titrated against 10.0 mL aliquots of $HCl_{(aq)}$ of unknown concentration:

**Table 16.12.** Volume of 0.200 mol/L $Ba(OH)_{2(aq)}$ used.

| Trial | 1 | 2 | 3 | 4 |
|---|---|---|---|---|
| Final burette reading (mL) | 12.4 | 24.7 | 37.1 | 49.5 |
| Initial burette reading (mL) | 0 | 12.4 | 24.7 | 37.1 |
| Volume of $Ba(OH)_{2(aq)}$ used (mL) | 12.4 | 12.3 | 12.4 | 12.4 |

*Provided:*

Volume of $Ba(OH)_{2(aq)}$ (to be calculated)

Concentration of $Ba(OH)_{2(aq)} = 0.200$ mol/L

$$\text{Volume of } HCl_{(aq)} = 10.0 \text{ mL} = \frac{10.0 \text{ mL}}{1000 \text{ mL/L}} = 0.0100 \text{ L}$$

*Required:*

The concentration of the $HCl_{(aq)}$

*Strategy:*

To solve this problem, first find the average volume of the $Ba(OH)_{2(aq)}$ this will allow number of moles of $Ba(OH)_{2(aq)}$ used in the titration to be calculated. Next balance an equation to show how $Ba(OH)_{2(aq)}$ and $HCl_{(aq)}$ react; this will allow number of moles of used $HCl_{(aq)}$ to be calculated. Knowing the number of moles and concentration of the $HCl_{(aq)}$, its volume can be calculated.

*Steps to be taken:*

**Step 1.** Find the average volume of the $Ba(OH)_{2(aq)}$ used (trial 2 is discarded because of inconsistency).

$$V = \frac{12.4\ mL + 12.4\ mL + 12.4\ mL}{3} = 12.4\ mL = \frac{12.4\ mL}{1000\ mL/L} = 0.0124\ L$$

**Step 2.** Find the number of moles of $Ba(OH)_{2(aq)}$ used.

Using the formula, $\mathbf{n = C \times V}$ we get

$$n = 0.200\ mol/L \times 0.0124\ L\ = \mathbf{2.84 \times 10^{-3}\ mol\ Ba(OH)_{2(aq)}}$$

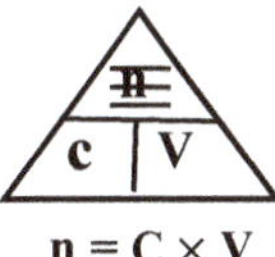


**Step 3.** Balance an equation to show how $Ba(OH)_{2(aq)}$ and $HCl_{(aq)}$ react

$$\underset{\text{Given}}{Ba(OH)_{2(aq}} + \underset{\text{Required}}{2\ HCl_{(aq)}} \rightarrow BaCl_{2(aq)} + 2\ H_2O_{(l)}$$

**Step 4.** Find the number of moles of $HCl_{(aq)}$ used.

$$\frac{Given}{Required} = \frac{1}{2} = \frac{2.84 \times 10^{-3}\ mol\ Ba(OH)_{2(aq)}}{X\ mol\ HCl(aq)}$$

$$X = \mathbf{5.68 \times 10^{-3}\ mol\ HCl_{(aq)}}$$

**Step 5.** Find the concentration of the $HCl_{(aq)}$.

Using the formula, $\mathbf{C = \dfrac{n}{V}}$ we get

$$\frac{n}{c \,|\, V} \quad C = \frac{n}{V}$$

$$C = \frac{5.68 \times 10^{-3}\ mol}{0.0100\ L} = \mathbf{0.568\ mol/L\ HCl_{(aq)}}$$

## Problem:

**c)** The table below shows the results obtained when a solution of $Ca(OH)_{2(aq)}$ of unknown concentration is titrated against 10.0 mL aliquots of 0.100 mol/L $HCl_{(aq)}$.

**Table 16.13.** Volume of $Ca(OH)_{2(aq)}$ of unknown concentration used.

| Trial | 1 | 2 | 3 | 4 |
|---|---|---|---|---|
| Final burette reading (mL) | 13.4 | 25.7 | 38.1 | 50.5 |
| Initial burette reading (mL) | 0 | 12.4 | 24.7 | 37.1 |
| Volume of $Ca(OH)_{2(aq)}$ used (mL) | 13.4 | 13.3 | 13.4 | 13.4 |

*Provided:*

Volume of $Ca(OH)_{2(aq)}$ (to be calculated)

Concentration of $HCl_{(aq)}$ = 0.100 mol/L

$$\text{Volume of } HCl_{(aq)} = 10.0\ mL = \frac{10.0\ mL}{1000.0\ mL/L} = 0.0100\ L$$

*Required:*

The concentration of the $Ca(OH)_{2(aq)}$

*Strategy:*

To solve this problem, first find the number of moles of the $HCl_{(aq)}$ used. Next, balance an equation to show how $Ca(OH)_{2(aq)}$ and $HCl_{(aq)}$ react; this will allow number of moles of $Ca(OH)_{2(aq)}$ used to be calculated. Knowing the number of moles and concentration of the $Ca(OH)_{2(aq)}$, its volume can be calculated.

*Steps to be taken:*

**Step 1.** Find the number of moles of the $HCl_{(aq)}$ used.

Using the formula, $\mathbf{n = C \times V}$ we get

$$\mathbf{n} = 0.10 \text{ mol/L} \times 0.0100 \text{ L} = \mathbf{0.0010 \text{ mol } HCl_{(aq)}}$$

**Step 2.** Find the average volume of the $Ca(OH)_{2(aq)}$ used (trial 2 is discarded because of inconsistency)

$$V = \frac{13.4 \text{ mL} + 13.4 \text{ mL} + 13.4 \text{ mL}}{3} = 13.3 \text{ mL} = \frac{13.4 \text{ mL}}{1000 \text{ mL/L}} = \mathbf{0.0134 \text{ L } Ca(OH)_{2(aq)}}$$

**Step 3.** Balance an equation to show how $Ca(OH)_{2(aq)}$ and $HCl_{(aq)}$ react.

|  | **Required** |  | **Given** |  |  |  |  |
| | | | | | | | |

$$\underset{\text{Required}}{Ca(OH)_{2(aq}} \quad + \quad \underset{\text{Given}}{2\,HCl_{(aq)}} \quad \rightarrow \quad BaCl_{2(aq)} \quad + \quad 2\,H_2O_{(l)}$$

**Step 4.** Find the number of moles of $Ca(OH)_{2(aq)}$ used.

$$\frac{\text{Given}}{\text{Required}} = \frac{2}{1} = \frac{0.0010 \text{ mol } HCl(aq)}{X \text{ mol } Ca(OH)_{2(aq)}}$$

$$\mathbf{2\,X = 0.0010 \text{ mol } Ca(OH)_{2(aq)} = X = 5.0 \times 10^{-4} \text{ mol } Ca(OH)_{2(aq)}}$$

**Step 5.** Find the concentration of the $Ca(OH)_{2(aq)}$.

Using the formula, $\mathbf{C = \dfrac{n}{V}}$ we get

$$\mathbf{C} = \frac{5.0 \times 10^{-4} \text{ mol}}{1.30 \times 10^{-2} \text{ L}} = \mathbf{3.8 \times 10^{-2} \text{ mol/L } Ca(OH)_{2(aq)}}$$

*Problem:*

**d)** The table below shows the results obtained when a solution of $Na_2CO_{3(aq)}$ of unknown concentration is titrated against 10.0 mL aliquots of 0.100 mol/L $H_2SO_{4(aq)}$

**Table 16.13.** Volume of $Na_2CO_{3(aq)}$ of unknown concentration used.

| Trial | 1 | 2 | 3 | 4 |
|---|---|---|---|---|
| Final burette reading (mL) | 15.2 | 30.3 | 45.5 | 50.7 |
| Initial burette reading (mL) | 0 | 15.2 | 30.3 | 35.5 |
| Volume of $Na_2CO_{3(aq)}$ used (mL) | 15.2 | 15.1 | 15.2 | 15.2 |

i) Use the evidence provided in the table to calculate the concentration of the $Na_2CO_{3(aq)}$.

ii) What mass of $Na_2CO_3$ must have been used to make 500.0 mL of this $Na_2CO_{3(aq)}$ solution?

*Provided:*

Volume of $Na_2CO_{3(aq)}$ (to be calculated)

Concentration of $H_2SO_{4(aq)}$ = 0.100 mol/L

Volume of $H_2SO_{4(aq)}$ = 10 mL = $\dfrac{10.0\ mL}{1000\ mL/L}$ = 0.0100L

*Required:*

The concentration of the $Na_2CO_{3(aq)}$

*Strategy:*

To solve this problem, first find the number of moles of the $H_2SO_{4(aq)}$ used. Next balance an equation to show how $H_2SO_{4(aq)}$ and $Na_2CO_{3(aq)}$ react; this will allow number of moles of $Na_2CO_{3(aq)}$ used to be calculated. Knowing the number of moles and volume of the $Na_2CO_{3(aq)}$, its concentration can be calculated.

For part ii), if the concentration of a certain volume of the $Na_2CO_{3(aq)}$ is known, its number of moles can be found. Knowing the number of moles, its mass can be calculated.

*Steps to taken:*

**Step 1.** Find the number of moles of used $H_2SO_{4(aq.)}$ Using the formula, $\mathbf{n = C \times V}$ we get

$$\mathbf{n = 0.100\ mol/L \times 0.0100\ L = 0.00100\ mol = 1.00 \times 10^{-3}\ mol\ H_2SO_{4(aq)}}$$

**Step 3.** Balance an equation to show how $H_2SO_{4(aq)}$ and $Na_2CO_{3(aq)}$ react.

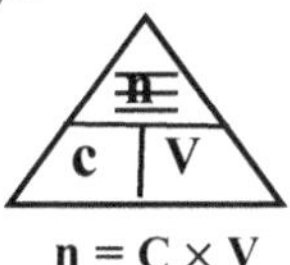


| Given | | Required | | | |
| --- | --- | --- | --- | --- | --- |
| $H_2SO_{4(aq)}$ | + | $Na_2CO_{3(aq)}$ | $\rightarrow$ | $Na_2SO_{4(aq)}$  +  $H_2O_{(l)}$  +  $CO_{2(g)}$ | |

**Step 4.** Find the number of moles of used $Na_2CO_{3(aq)}$.

$$\frac{\text{Given}}{\text{Required}} = \frac{1}{1} = \frac{1.00 \times 10^{-3}\ mol\ H_2SO_{4(aq)}}{X\ mol\ Na_2CO_{3(aq)}}$$

$$X = \mathbf{1.00 \times 10^{-3}\ mol\ Na_2CO_{3(aq)}}$$

**Step 5.** Find the average volume of $Na_2CO_{3(aq)}$ used (trial 2 discarded because of inconsistency)

$$V = \frac{15.2\ mL + 15.2\ mL + 15.2\ mL}{3} = 15.2\ mL = \frac{15.2\ mL}{1000\ mL/L} = \mathbf{0.0152\ L\ Na_2CO_{3(aq)}}$$

**Step 6.** Find the concentration of the $Na_2CO_{3(aq)}$

Using the formula, $\mathbf{C = \dfrac{n}{V}}$ we get

$$C = \frac{1.00 \times 10^{-3}\ mol}{0.0152\ L} = \mathbf{0.0657\ mol/L\ Na_2CO_{3(aq)}}$$

**Step 7.** Find the number of moles of $Na_2CO_{3(aq)}$ in 500.0 mL (0.5000 L) of 0.0657 mol/L.

Using the formula, $\mathbf{n = C \times V}$ we get

$$\mathbf{n = 0.0657\ mol/L \times 0.5000\ L = 0.0328\ mol\ Na_2CO_{3(aq)}}$$

**Step 8.** Find the mass of 0.034 mol $Na_2CO_{3(aq)}$. First find the molar mass of $Na_2CO_3$.

$$M = 2 \times 22.99\ g/mol\ Na + 1 \times 12.01\ g/mol\ C + 3 \times 16.00\ g/mol\ O = \mathbf{106\ g/mol\ Na_2CO_{3(s)}}$$

Using the formula, $\mathbf{m = n \times M}$ we get

Mass = 0.0328 mol × 106 g/mol = **3.48 g $Na_2CO_3$**. This the mass that must be used to make

500.0  mL of **0.0679 mol/L $Na_2CO_{3(aq)}$**

*Problem:*

  **a)** Calculate the pH of a $1.0\times10^{-3}$ mol/L $HNO_{3(aq)}$ solution.

*Provided:*

Concentration of $HNO_{3(aq)} = 1.0\times10^{-3}$ mol/L

- $HNO_{3(aq)}$ and all the other acids used are strong and they therefore ionize completely in water.

*Strategy:*

Show how the acid ionizes as this will help to find the $[H^+_{(aq)}]$ from which its its pH can be calculated.

*Steps to be  followed:*

**Step 1.** Complete the ionization equation to show the number of moles of $H^+_{(aq)}$ ions produced per mole of the acid as it ionizes.     $HNO_{3(aq)}$     $\rightarrow$     $H^+_{(aq)}$     +     $NO_3^-{}_{aq)}$

$\qquad\qquad\qquad$ 1 mol/L $\qquad\qquad$ 1 mol/L $\qquad\qquad$ 1 mol/L

**Step 2.**  Calculate the $[H^+_{(aq)}]$ of the solution.

$$[H^+_{(aq)}] = 1.0\times10^{-3} \text{ mol/L } HNO_{3(aq)} \times \frac{1 \text{ mol/L } H^+(aq)}{1\frac{mol}{L}HNO_3(aq)} = 1.0 \times10^{-3} \text{ mol/L } [H^+_{(aq)}]$$

**Step 3.** Using the formula, $pH = -\log [H^+_{(aq)}]$ we get

$$pH = -\log 1.0\times10^{-3} = \mathbf{3.0}$$

**Note that the following problems are similar to the one solved before, so for the sake of time and space we will use some short cuts.*

*Problem:*

**b)** Calculate the pH of a $1.2\times10^{-2}$ mol/L $HCl_{(aq)}$ solution.

*Steps to be followed:*

**Step 1.** Complete the ionization equation to show the number of moles of $H^+_{(aq)}$ ions produced per mole of the acid as it ionizes.     $HCl_{(aq)}$     $\rightarrow$     $H^+_{(aq)}$     +     $Cl^-_{aq)}$

$\qquad\qquad\qquad$ 1 mol/L $\qquad\qquad$ 1 mol/L $\qquad\qquad$ 1 mol/L

$\qquad\qquad\qquad$ **$1.2 \times 10^{-2}$ mol/L** $\qquad$ $1.2 \times 10^{-2}$ mol/L

**Step 2.** $pH = -\log 1.2 \times 10^{-2} = \mathbf{1.9}$

*Problem:*

**c)** Calculate the pH of a $2.4 \times 10^{-4}$ mol/L $HClO_{3(aq)}$ solution.

*Steps to be followed:*

**Step 1.** Complete the ionization equation to show the number of moles of $H^+_{(aq)}$ ions produced per mole of the acid as it ionizes.     $HClO_{3(aq)}$     $\rightarrow$     $H^+_{(aq)}$     +     $ClO_3^-{}_{(aq)}$

$\qquad\qquad\qquad$ 1 mol/L $\qquad\qquad$ 1 mol/L $\qquad\qquad$ 1 mol/L

$\qquad\qquad\qquad$ **$2.4 \times 10^{-4}$ mol/L** $\qquad$ $2.4 \times 10^{-4}$ mol/L

**Step 2.** Find the pH of the solution.

$$pH = -\log 2.4 \times 10^{-4} = \mathbf{3.6}$$

**Exercise 17.7 (p.295)**

***Problem:***

**a)** Determine the pH of a solution that has 6.30 g $HNO_{3(aq)}$ dissolved in 1.0 L of solution.

*Provided:*

Mass of $HNO_{3(aq)}$ = 6.30 g

Volume of solution = 1.0 L

*Required:*

pH of the solution

*Strategy:*

To solve this problem, first find the number of moles of the $HNO_3$ that are dissolved in the solution. Knowing this, the concentration of the solution can be calculated, from which the concentration of $H^+_{(aq)}$ and pH can subsequently be determined.

*Steps to be followed:*

**Step 1.** Find the number of moles of $HNO_{3(aq)}$ in 6.30 g. To do this, first find the molar mass of $HNO_3$.

$$M = 1 \times 1.01 \text{ g/mol H} + 1 \times 14.01 \text{ g/mol N} + 3 \times 16.00 \text{ g/mol O} = \mathbf{\textit{63.02 g/mol}}$$

Using the formula, $\mathbf{n} = \dfrac{m}{M}$ we get $\mathbf{n} = \dfrac{6.30g}{63.02 \text{ g/mol}} = \mathbf{0.100 \ mol \ HNO_3}$

**Step 2.** Find the concentration of $HNO_{3(aq)}$.

Using the formula, $\mathbf{C} = \dfrac{n}{V}$ we get, $\mathbf{C} = \dfrac{0.100 \text{ mol}}{1.0 \text{ L}} = \mathbf{0.10 \ mol/L \ HNO_3}$

**Step 3.** Find the concentration of $H^+_{(aq)}$ in $HNO_{3(aq)}$ the solution, using the ionization equation.

| . | $HNO_{3(aq)}$ | $\longrightarrow$ | $H^+_{(aq)}$ | + | $NO_3^-{}_{(aq)}$ |
|---|---|---|---|---|---|
| | 1 mol/L | | 1 mol/L | | |
| | **0.10 mol/L** | | **0.10 mol/L** | | |

**Step 4.** Find the pH of the solution

$$\mathbf{pH = -\log 0.10 = 1.0}$$

***Problem:***

**b)** Determine the pH of a solution that has 18.25 g $HCl_{(aq)}$ dissolved in 2.0 L of solution.

*Provided:*

Mass of $HCl_{(aq)}$ = 18.25 g

Volume of solution = 2.0 L

*Required:*

pH of the solution

*Strategy:*

To solve this problem, first find the number of moles of HCl$_{(aq)}$ the that are dissolved in the solution. Knowing this, the concentration of the solution can be calculated, from which the concentration of H$^+_{(aq)}$ and pH can subsequently be determined.

*Steps to be followed:*

**Step 1.** Find the number of moles of HCl$_{(aq)}$ in 6.30 g. To do this, first find the molar mass of HCl$_{(aq)}$.

M = 1 × 1.01 g/mol H + 1 × 35.45 g/mol Cl = **36.46 g/mol**

Using the formula, $\mathbf{n = \frac{m}{M}}$ we get $\mathbf{n = \dfrac{18.25g}{36.46 \text{ g/mol}}}$ = **0.500 mol HCl**

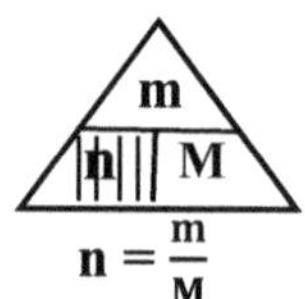

**Step 2.** Find the concentration of HCl$_{(aq)}$.

Using the formula, $\mathbf{C = \frac{n}{V}}$ we get $\mathbf{C = \dfrac{0.500 \text{ mol}}{2.0 \text{ L}}}$ = **0.25 mol/L HCl$_{(aq)}$**

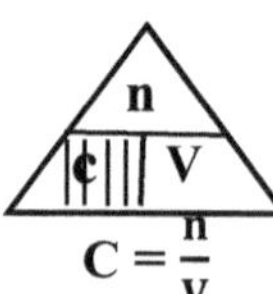

**Step 3.** Find the concentration of H$^+_{(aq)}$ in HCl$_{(aq)}$ the solution, using the ionization equation.

| . HCl$_{(aq)}$ → | H$^+_{(aq)}$ | + | Cl$^-_{(aq)}$ |
|---|---|---|---|
| 1 mol/L | 1 mol/L | | |
| **0.25 mol/L** | **0.25 mol/L** | | |

**Step 4.** Find the pH of the solution

pH = -log 0.25 = **0.60**

## Problem:

**c)** Determine the pH of a solution that has 8.45 g HClO$_{3(aq)}$ dissolved in 4.0 L of solution.

*Provided:*

Mass of HClO$_{3(aq)}$ = 8.45 g

Volume of solution = 4.0 L

*Required:*

pH of the solution

*Strategy:*

To solve this problem, first find the number of moles of the HClO$_{3(aq)}$ that are dissolved in the solution. Knowing this, the concentration of the solution can be calculated, from which the concentration of H$^+_{(aq)}$ and pH can subsequently be determined.

*Steps to be followed:*

**Step 1.** Find the number of moles of HClO$_{3(aq)}$ in 8.45 g. To do this, first find the molar mass of HClO$_{3(aq)}$.

M = 1 × 1.01 g/mol H + 1 × 35.45 g/mol Cl + 3 × 16.00 g/mol O = **84.46 g/mol**

Using the formula, $\mathbf{n = \frac{m}{M}}$ we get $\mathbf{n = \dfrac{8.45 \text{ g}}{84.46 \text{ g/mol}}}$ = **0.100 mol HClO$_{3(aq)}$**

**Step 2.** Find the concentration of $HCl_{(aq)}$.

Using the formula, $C = \dfrac{n}{V}$ we get $C = \dfrac{0.10 \text{ mol}}{4.0 \text{ L}} = $ **0.025 mol/L $HCl_{(aq)}$**

**Step 3.** Find the concentration of $H^+_{(aq)}$ in $HCl_{(aq)}$ the solution, using the ionization equation.

|   | $HClO_{3(aq)}$ | $\rightarrow$ | $H^+_{(aq)}$ | $+$ | $ClO_3^-{}_{(aq)}$ |
|---|---|---|---|---|---|
| . | 1 mol/L | | 1 mol/L | | |
| | **0.025 mol/L** | | **0.025 mol/L** | | |

**Step 4.** Find the pH of the solution

$$pH = -\log 0.025 = \textbf{0.16}$$

## *Problem:*

**d)** Determine the pH of a solution that has 1.28 g $HI_{(aq)}$ dissolved in 400.0 mL of solution.

*Provided:*

Mass of $HI_{(aq)} = 1.28$ g

Volume of solution $= 400.0 \text{ mL} = \dfrac{400.0 \text{ mL}}{1000 \text{ mL/L}} = 0.4000 \text{ L}$

*Required:*

pH of the solution

*Strategy:*

To solve this problem, first find the number of moles of the $HI_{(aq)}$ that are dissolved in the solution. Knowing this, the concentration of the solution can be calculated, from which the concentration of $H^+_{(aq)}$ and pH can subsequently, determined.

*Steps to be followed:*

**Step 1.** Find the number of moles of $HI_{(aq)}$ in 1.28 g. To do this, first find the molar mass of $HI_{(aq)}$.

$$M = 1 \times 1.01 \text{ g/mol H} + 1 \times 126.9 \text{ g/mol I} = \textit{128 g/mol}$$

Using the formula, $n = \dfrac{m}{M}$ we get $n = \dfrac{1.28\text{g}}{128 \text{ g/mol}} = $ **0.0100 mol $HI_{(aq)}$**

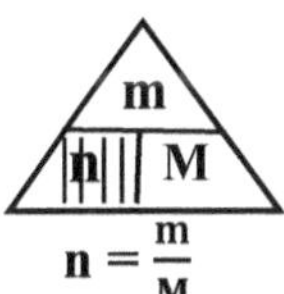

**Step 2.** Find the concentration of $HCl_{(aq)}$.

Using the formula, $C = \dfrac{n}{V}$ we get $C = \dfrac{0.010 \; mol}{0.400 \; L} = $ **0.025 mol/L $HI_{(aq)}$**

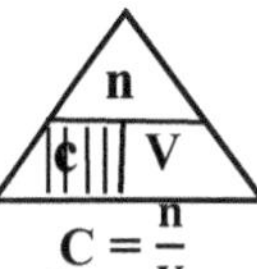

**Step 3.** Find the concentration of $H^+_{(aq)}$ in $HI_{(aq}$ the solution, using the ionization equation.

|   | $HI_{(aq)} \rightarrow$ | $H^+_{(aq)}$ | $+$ | $I^-_{(aq)}$ |
|---|---|---|---|---|
| . | 1 mol/L | 1 mol/L | | |
| | **0.025 mol/L** | **0.025 mol/L** | | |

**Step 4.** Find the pH of the solution

$$pH = -\log 0.025 = \textbf{1.6}$$

*Problem:*

**a)** Calculate the $[H^+_{(aq)}]$ in a solution of $H_2SO_{4(aq)}$ having a pH pf 2.0.

*Provided:*

pH of the solution = 2.0

*Required:*

The concentration of $H^+_{(aq)}$.

*Steps to be followed:*

**Step 1.** Using the equation $[H^+_{(aq)}] = 1 \times 10^{-pH}$, we get

$$[H^+_{(aq)}] = \mathbf{1.0 \times 10^{-2} \ mol/L}$$

*Problem:*

**b)** Calculate the $[H^+_{(aq)}]$ in a solution of $HCl_{(aq)}$ having a pH pf 4.0.

*Provided:*

pH of the solution = 4.0

**Step 1.** Using the equation $[H^+_{(aq)}] = 1 \times 10^{-pH}$, we get,

$$[H^+_{(aq)}] = \mathbf{1.0 \times 10^{-4} \ mol/L}$$

*Problem:*

**c)** Calculate the $[H^+_{(aq)}]$ in a solution of $HClO_{3(aq)}$ having a pH pf 4.4.

*Given:*

pH of the solution = 4.4

**Step 1.** Using the equation $[H^+_{(aq)}] = 1 \times 10^{-pH}$, we get,

$$[H^+_{(aq)}] = 1 \times 10^{-4.4} \ mol/L = \mathbf{3.98 \times 10^{-5} \ mol/L}$$

*Problem:*

**d)** Calculate the $[H^+_{(aq)}]$ in a solution of $HI_{(aq)}$ having a pH 3.0.

*Provided:*

pH of the solution = 3.0

**Step 1.** Using the equation $[H^+_{(aq)}] = 1 \times 10^{-pH}$, we get

$$[H^+_{(aq)}] = \mathbf{1.0 \times 10^{-3} \ mol/L}$$

*Problem:*

**e)** Calculate the $[H^+_{(aq)}]$ in a solution of $CH_3COOH_{(aq)}$ of pH 5.0.

*Given:*

pH of the solution = 5.0

**Step 1.** Using the equation $[H^+_{(aq)}] = 1 \times 10^{-pH}$, we get,

$$[H^+_{(aq)}] = \mathbf{1 \times 10^{-5} \ mol/L}$$

*Problem:*

**a).** A solution of $HCl_{(aq)}$ of pH 2.0 is diluted until its new pH is 5.0. What is the change in $[H^+_{(aq)}]$?

*Provided:*

Original pH of $HCl_{(aq)}$ = 2.0

Final pH of $HCl_{(aq)}$ = 5.0

*Required:*

The change in $[H^+_{(aq)}]$

*Strategy:*

To sovle this problem, change pH to $[H^+_{(aq)}]$ in both instances then compare their change in concentration.

*Steps to be followed:*

**Step 1.** Change the original and final pH to $[H^+_{(aq)}]$, using the formula, $[H^+_{(aq)}] = 1 \times 10^{-pH}$

Original $[H^+_{(aq)}] = 1 \times 10^{-2}$ mol/L = **0.010 mol/L**

Final $[H^+_{(aq)}] = 1 \times 10^{-5}$ mol/L = **0.000010 mol/L**

*The final concentration is **1000 times less** than the original*

**Problem:**

**b).** A solution of $HNO_{3(aq)}$ of pH 1.0 is diluted until its new pH is 6.0. What is the change in $[H^+_{(aq)}]$?

*Provided:*

Original pH of $HNO_{3(aq)}$ = 1.0

Final pH of $HNO_{3(aq)}$ = 6.0

*Required:*

The change in $[H^+_{(aq)}]$

*Strategy:*

To sovle this problem, change pH to $[H^+_{(aq)}]$ in both instances then compare their change in concentration.

*Steps to be followed:*

**Step 1.** Change the original and final pH to $[H^+_{(aq)}]$, using the formula, $[H^+_{(aq)}] = 1 \times 10^{-pH}$

Original $[H^+_{(aq)}] = 1 \times 10^{-1}$ mol/L = **0.10 mol/L**

Final $[H^+_{(aq)}] = 1 \times 10^{-6}$ mol/L = **0.0000010 mol/L**

*The final concentration is **10 000 times** less than the original*

**Problem:**

**c)** To what new volume must a 100.0 mL of $1.0 \times 10^{-2}$ mol/L $HCl_{(aq)}$ be diluted to achieve a final pH of 4.0?

*Provided:*

Original concentration of $HCl_{(aq)} = 1.0 \times 10^{-2}$ mol/L

Original volume, $(V_1)$ of $HCl_{(aq)} = 100.0$ mL $= \dfrac{100.0\,\text{mL}}{1000\ \text{mL/L}} = 0.1000$ L

Final pH $= 4.0$

*Required*: Final volume?

*Strategy:*

Knowing the original concentration of $HCl_{(aq)}$, its $[H^+_{(aq)}]$ can calculated. Also, the final pH is also given and this it can be used to calculate the final $[H^+_{(aq)}]$. Since the initial volume is provided, the final volume can be found.

*Steps to be followed:*

**Step 1.** Find the original $[H^+_{(aq)}]$ $(C_1)$ that is produced from the ionization of $1.0 \times 10^{-2}$ mol/L $HCl_{(aq)}$.

$$HCl_{(aq)} \quad \longrightarrow \quad H^+_{(aq)} \quad + \quad Cl^-_{(aq)}$$

$$1\ \text{mol/L} \qquad\qquad 1\ \text{mol/L}$$

$$\mathbf{1.0 \times 10^{-2}\ mol/L} \qquad \mathbf{1.0 \times 10^{-2}\ mol/L}$$

**Step 2.** Find the final $[H^+_{(aq)}]$, $(C_1)$ that produced a pH of 4.0

Using the equation $[H^+_{(aq)}] = 1 \times 10^{-pH}$ we get

$$[H^+_{(aq)}] = \mathbf{1.0 \times 10^{-4}\ mol/L}$$

**Step 3.** Find the final volume, $(V_2)$ to which the acid has been diluted.

Using the formula, $C_1 \times V_1 = C_2 \times V_2$ and making $V_2$ the subject of the equation and   substituting values we get

$$V_2 = \frac{C_1 \times V_1}{C_2} = \frac{1.0 \times 10^{-2}\,\text{mol/L} \times 1.00 \times 10^{-1}\,\text{L}}{1.0 \times 10^{-4}\,\text{mol/L}} = \mathbf{1.0 \times 10^{1}\ L\ HCl_{(aq)}}$$

**Problem:**

**d)** A solution is having 12.79 g of $HI_{(aq)}$ dissolved in 2.0 L is diluted to 4.0 L. What is its

      i) Original pH?

      ii) Final pH?

      ii) Change in $[H^+_{(aq)}]$

*Provided:*

Mass of HI dissolved in 2.0 L $= 12.79$ g

Original volume, $(V_1)$ of $HI_{(aq)} = 2.0$ L
Final volume, $(V_2)$ of $HI_{(aq)} \quad = 4.0$ L

*Required:*

Original and final pH

Change in $[H^+_{(aq)}]$

*Strategy:*

To solve this problem, first find the number of moles of the $HI_{(aq)}$ that are dissolved in 2.0 L of solution. Knowing this, the original concentration, ($C_1$) of the solution can be calculated, from which the concentration of $H^+_{(aq)}$ and pH can be subsequently determined. Knowing the final volume of 4.0 L, the final concentration, ($C_2$) of $H^+_{(aq)}$ and pH can be finally determined.

*Steps to be followed*

**Step 1.** Find the number of moles of HI in 12.79 g.

First find the molar mass of HI.

$M = 1 \times 1.01$ g/mol H $+ 1 \times 126.9$ g/mol I $= \boldsymbol{128 \ g/mol}$

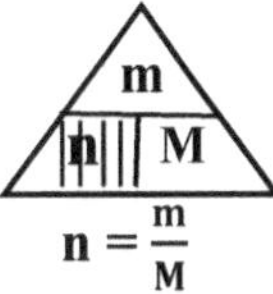

Using the formula, $\mathbf{n = \dfrac{m}{M}}$ we get $\mathbf{n = \dfrac{12.79 \ g}{128 \ g/mol} = 0.1000 \ mol \ HI}$

**Step 2.** Find the concentration of $HI_{(aq)}$.

Using the formula, $\mathbf{C = \dfrac{n}{V}}$ we get $\mathbf{C = \dfrac{0.1000 \ mol}{2.0 \ L} = 0.050 \ mol/L \ HI_{(aq)}}$

**Step 3.** Find the concentration of $H^+_{(aq)}$ in $HI_{(aq}$ the solution, using the ionization equation.

$$HI_{(aq)} \quad \longrightarrow \quad H^+_{(aq)} \quad + \quad I^-_{(aq)}$$

$$1 \ mol/L \qquad \qquad 1 \ mol/L$$

$$0.050 \ mol/L \qquad \qquad \boldsymbol{0.050 \ mol/L}$$

**Step 4.** Find the original pH of the solution.

pH = -log 0.050 = **1.3**

**Step 5.** Find the final concentration ($C_2$) of the diluted acid.

Using the formula, $C_1 \times V_1 = C_2 \times V_2$ and making $C_2$ the subject of the equation and substituting values, we get

$$C_2 = \frac{C_1 \times V_1}{V_2} = \frac{0.050 \ mol/L \times 2.0 \ L}{4.0 L} = \boldsymbol{0.025 \ mol/L \ HI_{(aq)}}$$

**Step 6.** Find the final $[H^+_{(aq)}]$.   $HI_{(aq)} \quad \longrightarrow \quad H^+_{(aq)} \quad + \quad I^-_{(aq)}$

$$\boldsymbol{0.025 \ mol/L} \qquad \boldsymbol{0.025 \ mol/L}$$

**Step 7.** Find the final pH of the solution.

pH = -log 0.025 = **1.6**

*The final $[H^+_{(aq)}]$* **is 2 times less** *concentrated than the original*

## *Problem:*

**e)** If a solution has 2.0 g of $HCl_{(aq)}$ dissolved in 1.0 L of solution, determine the following:

i) Its pOH

ii) Its pH

*Provided:*

Mass of HCl dissolved in 1.0 L = 2.0 g

  Volume, of $HCl_{(aq)}$ = 1.0 L

*Required:*

  pOH and pH of the solution

*Strategy:*

To solve this problem, first find the number of moles of the $HCl_{(aq)}$ that are dissolved in 1.0 L of solution. Knowing this concentration, the concentration of $H^+_{(aq)}$ and pH can be determined. Knowing pH, the pOH can be subsequently determined.

*Steps to be followed:*

**Step 1.** Find the number of moles of $HCl_{(aq)}$ in 2.0 g. To do this, first find its molar mass.

    M = 1 × 1.01 g/mol H + 1 × 35.45 g/mol Cl = *36.46 g/mol*

Using the formula, $\mathbf{n = \dfrac{m}{M}}$ we get $\mathbf{n = \dfrac{2.0\ g}{36.46\ g/mol}}$ = **0.054 mol HCl**

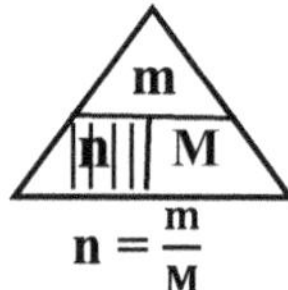

**Step 2.** Find the concentration of $HCl_{(aq)}$

  Using the formula, $\mathbf{C = \dfrac{n}{V}}$ we get $\mathbf{C = \dfrac{0.050\ mol}{1.0\ L}}$ = **0.054 mol/L $HCl_{(aq)}$**

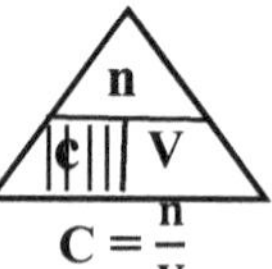

**Step 3.** Find the concentration of $H^+_{(aq)}$ in $HCl_{(aq)}$ the solution, using the ionization equation.

$$. \quad HCl_{(aq)} \quad \longrightarrow \quad H^+_{(aq)} \quad + \quad Cl^-_{(aq)}$$

        **0.054 mol/L**       **0.054 mol/L**

**Step 4.** Find the pH of the solution.

      pH = -log 0.054 = **1.2**

 **Step 5.** Find the pOH of the solution.

Using the formula, pH + pOH = 14.0 we get **pOH** = 14.0 − 1.2 = **12.8**

***Problem:***

**f)** What mass of $NaOH_{(s)}$ is required to neutralize a 200.0 mL $HCl_{(aq)}$ of pH 2.0?

*Provided:*

  Volume of $HCl_{(aq)}$ = 200.0 mL = $\dfrac{200.0\ mL}{1000\ mL/L}$ = 0.2000 L

*Required:*

  pH of the solution

  The mass of $NaOH_{(s)}$ is required to neutralize a 200.0 mL $HCl_{(aq)}$ of pH 2.0.

*Strategy:*

To solve this problem, use the pH to find the $[H^+_{(aq)}]$ of the solution, from which the concentration of the $HCl_{(aq)}$ can be determined. Knowing the concentration and volume of the $HCl_{(aq)}$, its number of moles can be calculated. Using the mole ratios in the balanced equation between the reaction of $HCl_{(aq)}$ and $NaOH_{(s)}$, the number of moles of $NaOH_{(s)}$ can be found from which its mass can be calculated.

*Steps to be followed:*

**Step 1.** Convert pH of the solution to $[H^+_{(aq)}]$.

$[H^+_{(aq)}] = 1.0 \times 10^{-pH} = 1.0 \times 10^{-2}$ mol/L

**Step 2.** Find the concentration of $HCl_{(aq)}$ that produced $1.0 \times 10^{-2}$ mol/L $H^+_{(aq)}$.

$$HCl_{(aq)} \quad \rightarrow \quad H^+_{(aq)} \quad + \quad Cl^-_{(aq)}$$
1 mol /L $\qquad$ 1 mol/L
**$1.0 \times 10^{-2}$ mol/L** $\qquad$ **$1.0 \times 10^{-2}$ mol/L**

Since 1 mol/L of $H^+_{(aq)}$ is produced from 1 mol/L of $HCl_{(aq)}$, then $1.0 \times 10^{-2}$ mol/L of it must have come from $1.0 \times 10^{-2}$ mol/L $HCl_{(aq)}$.

**Step 3.** Find the number $HCl_{(aq)}$ in 200.0 mL (0.200L) of the solution.

Using the formula, **n = C × V** we get

**n** = $1.0 \times 10^{-2}$ mol/L × 0.2000 L = **0.0020 mol HCl$_{(aq)}$**

**Step 4.** Balance an equation to show the reaction between $NaOH_{(aq)}$ and $HCl_{(aq)}$. In this case, the $HCl_{(aq)}$ becomes *the given reagent* and the NaOH *required one*.

**Required** $\qquad$ **Given**
$NaOH_{(aq)}$ $\quad + \quad HCl_{(aq)} \quad \rightarrow \quad NaCl_{(aq)} \quad + \quad H_2O_{(l)}$

**Step 5.** Find the number of moles of $NaOH_{(aq)}$ required to react with 0.002 moles of $HCl_{(aq)}$

$$\frac{Given}{Required} = \frac{1}{1} = \frac{0.0020 \text{ mol } HCl_{(aq)}}{X \text{ mol } NaOH_{(aq)}}$$

**X = 0.0020 mol NaOH$_{(aq)}$**

**Step 6.** Find the mass of 0.0020 mol $NaOH_{(aq)}$

Using the formula, **m = n × M**, we get

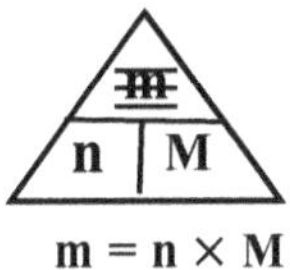

Mass = 0.0020 mol × 40.0 g/mol = **0.080 g NaOH$_{(aq)}$**

## Problem:

**g)** If 200.0 mL of 0.10 mol/L $HCl_{(aq)}$ is mixed with 300.0 mL of 0.20 mol/L $NaOH_{(aq)}$. Determine the pH of the resulting mixture.

*Provided:*

Volume of $HCl_{(aq)}$ = 200.0 mL = $\dfrac{200.0 \text{ mL}}{1000 \text{ mL/L}}$ = 0.2000 L $HCl_{(aq)}$

Concentration of $HCl_{(aq)}$ = 0.10 mol/L

Volume of $NaOH_{(aq)}$ = 300.0 mL = $\dfrac{300.0 \text{ mL}}{1000 \text{ mL/L}}$ = 0.3000 L

Concentration of $NaOH_{(aq)}$ = 0.20 mol/L

*Required:*

The pH of the mixture resulting from the reacting between 200.0 mL of 0.10 mol/L $HCl_{(aq)}$ and 300.0 mL of 0.20 mol/L $NaOH_{(aq)}$.

*Strategy:*

This problem entails finding which of the two reagents is in excess. Knowing this, the final pH of the mixture can be found. First find the number of moles of $HCl_{(aq)}$ and $NaOH_{(aq)}$ added; this would enable the amount of the excess reagent to be found. Knowing the number of moles of the excess reagent, its final concentration can be calculated since the final volume of the mixture is known; the sum of the volumes of the $HCl_{(aq)}$ and $NaOH_{(aq)}$ added.

*Steps to be followed:*

**Step 1**. Find the number of moles of the $HCl_{(aq)}$.

Using the equation, $\mathbf{n = C \times V}$, we get

$\mathbf{n}$ = 0.10 mol/L × 0.200 L = **0.020 mol HCl$_{(aq)}$ (Used)**

**Step 2**. Find the number of moles of the $NaOH_{(aq)}$.

Using the equation, $\mathbf{n = C \times V}$, we get

$\mathbf{n}$ = 0.30 mol/L × 0.200 L = **0.060 mol NaOH$_{(aq)}$ (Used)**

**Step 3**. Find the excess reagent.

To do this first balance an equation to show the reaction between $NaOH_{(aq)}$ and $HCl_{(aq)}$, then choose one of the reagent to find how of the other one is required to react with it; the **HCl$_{(aq)}$ is chosen** so becomes the *given reagent* and the NaOH *required one*.

<table>
<tr><td>**Required**</td><td></td><td>**Given**</td><td></td><td></td><td></td><td></td></tr>
<tr><td>$NaOH_{(aq)}$</td><td>+</td><td>$HCl_{(aq)}$</td><td>→</td><td>$NaCl_{(aq)}$</td><td>+</td><td>$H_2O_{(l)}$</td></tr>
</table>

**Step 5**. Find the number of moles of $NaOH_{(aq)}$ required to react with 0.002 moles of $HCl_{(aq)}$

$$\frac{\text{Given}}{\text{Required}} = \frac{1}{1} = \frac{0.020 \text{ mol } HCl_{(aq)}}{X \text{ mol NaOH(aq)}} \qquad X = \mathbf{0.020 \text{ mol NaOH}_{(aq)} \text{ (Required)}}$$

Based on the calculations, it can be seen that **0.020mol NaOH$_{(aq)}$** was required to react with the $HCl_{(aq)}$, but **0.060 mol NaOH$_{(aq)}$** was added. This means that there was an excess of **0.040 mol NaOH$_{(aq)}$** in the mixture.

**Step 6**. Find the concentration of the **excess NaOH$_{(aq)}$ (0.040 mol)** in the mixture. Note that the volume that the excess 0.040 mol NaOH$_{(aq)}$ is now dissolved in is **0.50 L, (0.20 L** from the $HCl_{(aq)}$ and **0.30L** from the $NaOH_{(aq)}$).

Using the formula, $\mathbf{C = \dfrac{n}{V}}$ we get $C = \dfrac{0.040 \text{ mol}}{0.50 \text{ L}} = \mathbf{0.080 \text{ mol/L NaOH}_{(aq)}}$

**Step 7**. Find the concentration of the $[OH_{(aq)}]$ in the mixture. Show how $NaOH_{(aq)}$ ionizes.

<table>
<tr><td>$NaOH_{(aq)}$</td><td>→</td><td>$Na^+_{(aq)}$</td><td>+</td><td>$OH^-_{(aq)}$</td></tr>
<tr><td>1 mol/L</td><td></td><td></td><td></td><td>1 mol/L</td></tr>
<tr><td>**0.080 mol/L**</td><td></td><td></td><td></td><td>**0.080mol/L**</td></tr>
</table>

**Step 8.** Find the $[H^+_{(aq)}]$ of the solution, using the formula

Using the formula, $[OH_{(aq)}] \times [H^+_{(aq)}] = 1 \times 10^{-14}$ and making $[H^+_{(aq)}]$ the subject of the equation. Doing this we get

$$[H^+_{(aq)}] = \frac{1 \times 10^{-14}}{[OH^-]} = \frac{1 \times 10^{-14}}{8.0 \times 10^{-2}} = \mathbf{1.2 \times 10^{-13} \text{ mol/L}}$$

$$pH = -\log 1.2 \times 10^{-13} = \mathbf{13}$$

## *Problem:*

**h)** 200.0 mL of 0.10 mol/L $HClO_{3(aq)}$ is mixed with 100.0 mL of 0.10 mol/L $NaOH_{(aq)}$. Determine the pH of the resulting mixture.

*Provided:*

Volume of $HClO_{3(aq)}$ = 200.0 mL = $\dfrac{200.0\text{mL}}{1000\text{ mL/L}}$ = 0.2000 L

Concentration of $HClO_{3(aq)}$ = 0.10 mol/L

Volume of $NaOH_{(aq)}$ = 100.0 mL = $\dfrac{100.0\text{mL}}{1000\text{ mL/L}}$ = 0.1000 L

Concentration of $NaOH_{(aq)}$ = 0.10 mol/L

*Required:*

The pH of the mixture resulting from the reacting between 200.0 mL of 0.10 mol/L $HClO_{3(aq)}$ and 100.0 mL of 0.10 mol/L $NaOH_{(aq)}$.

*Strategy:*

This problem entails finding which of the two reagents is in excess. Knowing this, the final pH of the mixture can be found. First the number of moles of $HClO_{3(aq)}$ and $NaOH_{(aq)}$ added; this would enable the amount of the excess reagent to be found. Knowing the number of moles of the excess reagent, its final concentration can be calculated since the final volume of the mixture is known; the sum of the volumes of the $HClO_{3(aq)}$ and $NaOH_{(aq)}$ added.

*Steps to be followed:*

**Step 1.** Find the number of moles of the $HCl_{(aq)}$.

Using the equation, **n = C × V**, we get

**n = 0.10 mol/L × 0.200 L = 0.020 mol $HClO_{3(aq)}$ (Used)**

**Step 2.** Find the number of moles of the $NaOH_{(aq)}$.

Using the equation, **n = C × V**, we get

**n = 0.10 mol/L × 0.100 L = 0.010 mol $NaOH_{(aq)}$ (Used)**

$$n = C \times V$$

**Step 3.** Find the excess reagent. To do this, first balance an equation to show the reaction between $NaOH_{(aq)}$ and $HClO_{3(aq)}$ then choose one of the reagents to find how of the other one is required to react with it. The $HClO_{3(aq)}$ is chosen, so becomes the *given reagent* and the NaOH *required* one.

| **Required** | | **Given** | | | |
|---|---|---|---|---|---|
| $NaOH_{(aq)}$ | + | $HClO_{3(aq)}$ | $\rightarrow$ | $NaClO_{3(aq)}$ | + $H_2O_{(l)}$ |

**Step 5.** Find the number of moles of $NaOH_{(aq)}$ required to react with 0.020 mol $HClO_{3(aq)}$.

$$\dfrac{\text{Given}}{\text{Required}} = \dfrac{1}{1} = \dfrac{0.020 \text{ mol } HClO_{3\,(aq)}}{X \text{ mol NaOH(aq)}}$$

**X = 0.020 mol $NaOH_{(aq)}$ (Required)**

**0.010 mol $NaOH_{(aq)}$ (Used)**

Based on the calculations, it can be seen that **0.020 mole $NaOH_{(aq)}$** was required to react with the 0.020 mole $HClO_{3(aq)}$ **but only 0.010 mole $NaOH_{(aq)}$** was added. This means that there was a shortage of **0.010 mole $NaOH_{(aq)}$** to neutralize all the $HClO_{3(aq)}$. There was thus **an excess of 0.010 mole $HClO_{3(aq)}$** in the mixture.

**Step 6.** Find the concentration of the $HClO_{3(aq)}$ in the mixture. Note that the **final volume** that the excess 0.010 mol $HClO_{3(aq)}$ is now dissolved in is **0.30 L**; (**0.20 L** from the $HClO_{3(aq)}$ and **0.10 L** from the $NaOH_{(aq)}$).

Using the formula, $\mathbf{C = \dfrac{n}{V}}$ we get $C = \dfrac{0.010 \text{ mol}}{0.30 \text{ L}}$ = **0.033 mol/L $HClO_{3(aq)}$.**

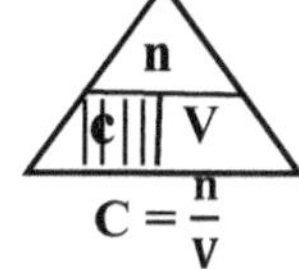

$$C = \dfrac{n}{V}$$

**Step 7.** Find the concentration of the $H^+_{(aq)}$ in the mixture. Show how $HClO_{3(aq)}$ ionizes.

$$HClO_{3(aq)} \rightarrow H^+_{(aq)} + ClO_3^-$$

$$\begin{array}{ccc} \text{1 mol/L} & & \text{1 mol/L} \\ \textbf{0.033 mol/L} & & \textbf{0.033 mol/L} \end{array}$$

**Step 8.** Find the pH of the solution with **0.033 mol/L** $H^+_{(aq)}$.

$$pH = -\log 3.3 \times 10^{-2} = \textbf{1.5}$$

## *Problem:*

**i)** If 2.0 g of $Mg_{(s)}$ is added to 200.0 mL of 1.0 mol/L $HCl_{(aq)}$, determine what volume of 1.0 mol/L $NaOH_{(aq)}$ is required to neutralize any excess acid.

*Provided:*

Mass of $Mg_{(s)}$ = 2.0 g

$$\text{Volume of } HCl_{(aq)} = 200.0 \text{ mL} = \frac{200.0 \text{mL}}{1000 \text{ mL/L}} = 0.2000 \text{ L}$$

Concentration of $HCl_{(aq)}$ = 1.0 mol/L
Concentration of $NaOH_{(aq)}$ = 0.10 mol/L

*Required:*

The volume of 1.0 mol/L $NaOH_{(aq)}$ is required to neutralize any excess $HCl_{(aq)}$.

*Strategy:*

This problem entails finding the amount of excess $HCl_{(aq)}$. Knowing this, the number of moles of $NaOH_{(aq)}$ to neutralize it can be calculated. First calculate the number of moles of $HCl_{(aq)}$ and $Mg_{(s)}$ and added; this would enable the amount of the excess $HCl_{(aq)}$ to be found.

*Steps to be followed:*

**Step 1.** Find the number of moles of the $HCl_{(aq)}$.

Using the equation, $\mathbf{n = C \times V}$, we get

$\mathbf{n = 1.0 \text{ mol/L} \times 0.200 \text{ L} = 0.20 \text{ mol } HCl_{(aq))} \textbf{ (Used)}}$

$$n = C \times V$$

**Step 2.** Find the number of moles of the $Mg_{(s)}$ added.

Using the formula, $\mathbf{n = \dfrac{m}{M}}$ we get $n = \dfrac{2.0 \text{ g}}{24.31 \text{ g/mol}} = \textbf{0.082 mol } Mg_{(s)}$

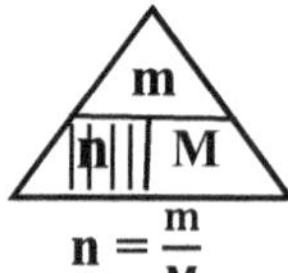

$$n = \frac{m}{M}$$

**Step 3.** Find the amount of excess $HCl_{(aq)}$.

To do this first balance an equation to show the reaction between and $HCl_{(aq)}$ then choose the $Mg_{(s)}$ as *the limiting reagent* to find how of the $HCl_{(aq)}$ is required to react with it.

$$\begin{array}{ccccccc} \textbf{Given} & & \textbf{Required} & & & & \\ Mg_{(s)} & + & 2\,HCl_{(aq)} & \rightarrow & MgCl_{2(aq)} & + & H_{2(g)} \end{array}$$

**Step 5.** Find the number of moles $HCl_{(aq)}$ required to react with **0.082 mol $Mg_{(s)}$.**

$$\frac{\text{Given}}{\text{Required}} = \frac{1}{2} = \frac{0.080 \text{ mol Mg}}{X \text{ mol } HCl_{(aq)}}$$

$$X = \textbf{0.16 mol } HCl_{(aq)} \textbf{ (Required)}$$

$$\textbf{0.20 mol } HCl_{(aq))} \textbf{ (Used)}$$

Based on the calculations, it can be seen that **0.16 mole HCl$_{(aq)}$** was required to react with the **0.082 mole Mg$_{(s)}$**. However, the amount of HCl$_{(aq)}$ used was **0.20 mole**. This means that there was an excess of **0.040 mole HCl$_{(aq)}$**.

**Step 6.** Find the number of moles of NaOH$_{(aq)}$ required to neutralize **0.040 mole of HCl$_{(aq)}$**.

First balance an equation to show how NaOH$_{(aq)}$ and HCl$_{(aq)}$ react.

$$\underset{\text{Required}}{\text{NaOH}_{(aq)}} \quad + \quad \underset{\text{Given}}{\text{HCl}_{(aq)}} \quad \rightarrow \quad \text{NaCl}_{(aq)} \quad + \quad \text{H}_2\text{O}_{(l)}$$

$$\frac{\text{Given}}{\text{Required}} = \frac{1}{1} = \frac{0.040 \text{ mol HCl}_{(aq)}}{\text{X mol NaOH(aq)}} \quad X = \textbf{0.040 mol NaOH}_{(aq)}$$

**Step 8.** Find the volume of the 0.10 mol/L NaOH$_{(aq)}$

$$\text{Using the formula, } \mathbf{V} = \frac{\mathbf{n}}{\mathbf{C}}, \text{ we get } \frac{0.040 \text{ mol}}{0.10 \text{ mol/L}} = \textbf{0.40 L} = \mathbf{4.0 \times 10^2 \text{ mL NaOH}_{(aq)}}$$

$$V = \frac{n}{C}$$

## *Problem:*

**j)** Only 200.0 mL of 0.50 mol/L Na$_2$CO$_{3(aq)}$ was available and used to neutralize 500.0 ml of 0.50 mol/L HCl$_{(aq)}$ spill. It was found that there was still an excess of the acid. Determine what mass of Na$_2$CO$_{3(s)}$ powder must be used to react with it for complete neutralization.

*Provided:*

$$\text{Volume of HCl}_{(aq)} = 500.0 \text{ mL} = \frac{500.0 \text{mL}}{1000 \text{ mL/L}} = 0.5000 \text{ L}$$

Concentration of HCl$_{(aq)}$ = 0.50 mol/L

Concentration of Na$_2$CO$_{3(aq)}$ = 0.50 mol/L

$$\text{Volume of Na}_2\text{CO}_{3(aq)} = 200.0 \text{ mL} = \frac{200.0 \text{mL}}{1000 \text{ mL/L}} = 0.2000 \text{ L}$$

*Required:*

The mass of Na$_2$CO$_{3(s)}$ powder must be used to neutralize any excess HCl$_{(aq)}$.

*Strategy:*

This problem entails finding the amount of excess HCl$_{(aq)}$ in the mixture. Knowing this, the number of moles of Na$_2$CO$_{3(s)}$ required to neutralize it can be calculated. First calculate the number of moles of HCl$_{(aq)}$ and Na$_2$CO$_{3(aq)}$ added; this would enable the amount of the excess HCl$_{(aq)}$ to be found. The amount of Na$_2$CO$_{3(s)}$ required to react completely with this can then be found.

*Steps to be followed:*

**Step 1.** Find the number of moles of the HCl$_{(aq)}$ used.

Using the equation, **n = C × V**, we get

**n = 0.50 mol/L × 0.500 L = 0.25 mol HCl$_{(aq)}$) (Used)**

**Step 2.** Find the number of moles of the Na$_2$CO$_{3(aq)}$ added.

Using the equation, **n = C × V**, we get

**n = 0.50 mol/L × 0.200 L = 0.10 mol Na$_2$CO$_{3(aq)}$ (Used)**

$$n = C \times V$$

**Step 3.** Find the amount of any excess reagent.

To do this, first balance an equation to show the reaction between Na$_2$CO$_{3(aq)}$ and HCl$_{(aq)}$, then choose **0.10 mol** Na$_2$CO$_{3(aq)}$ to find how of the HCl$_{(aq)}$ is required to react with it.

$$\underset{\text{Given}}{Na_2CO_{3(aq)}} + 2\,HCl_{(aq)} \rightarrow 2\,NaCl_{(aq)} + H_2O_{(l)} + CO_{2(g)}..$$

$$\frac{Given}{Required} = \frac{1}{2} = \frac{0.10 \text{ mol } Na_2CO_{3(aq)}}{X \text{ mol } HCl_{(aq)}}$$

$$X = \textbf{0.20 mol } HCl_{(aq)} \textbf{ (Required)}$$
$$= \textbf{0.25 mol } HCl_{(aq))} \textbf{ (Used)}$$

Based on the calculations, it can be seen that **0.20 mole** $HCl_{(aq)}$ was required to react **with the 0.10 mole** $Na_2CO_{3(aq)}$, but **0.25 mole** was used.

**Step 4.** Find the amount of excess $HCl_{(aq)}$ used.

$$= 0.25 \text{ mol} - 0.20 \text{ mol} = \textbf{0.050 mol } HCl_{(aq)}$$

**Step 5.** Find the number of moles of required to neutralize **0.050 mole of** $HCl_{(aq)}$.

First balance an equation to show how $Na_2CO_{3(s)}$ and $HCl_{(aq)}$ react.

$$\underset{\text{Required}}{Na_2CO_{3(s)}} + \underset{\text{Given}}{2\,HCl_{(aq)}} \rightarrow 2\,NaCl_{(aq)} + H_2O_{(l)} + CO_{2(g)}$$

$$\frac{Given}{Required} = \frac{2}{1} = \frac{0.050 \text{ mol } HCl_{(aq)}}{X \text{ mol } \text{ mol } Na_2CO_{3(s)}}$$

$$2\,X = \textbf{0.050 mol } Na_2CO_{3(s)}$$

$$X = \textbf{0.025 mol } Na_2CO_{3(s)}$$

**Step 8.** Find the mass of the 0.03 mol $Na_2CO_{3(s)}$. First find its molar mass.

$$M = 2 \times 22.99 \text{ g/mol Na} + 1 \times 12.01 \text{ g/mol C} + 3 \times 16.00 \text{ g/mol O} = \textbf{\textit{106 g/mol}}$$

Using the formula, $\mathbf{m = n \times M}$ we get

Mass $= 0.025 \text{ mol} \times 106 \text{ g/mol} = \textbf{2.6 g } Na_2CO_{3(s)}$

$$m = n \times M$$

## *Problem:*

k) During a titration, 2.00 mL of 0.10 mol/L of $H_2SO_{4(aq)}$ was added to a 10.0 mL aliquot of 0.10 mol/L $KOH_{(aq)}$ in an Erlenmeyer flask. Calculate the final pH of the mixture in the flask.

*Provided:*

Volume of $H_2SO_{4(aq)} = 2.0 \text{ mL} = \dfrac{2.0\text{mL}}{1000 \text{ mL/L}} = 0.0020 \text{ L} = 2.0 \times 10^{-3} \text{ L}$

Concentration of $H_2SO_{4(aq)} = 0.10$ mol/L

Volume of $KOH_{(aq)} = 10.0 \text{ mL} = \dfrac{10.0\text{mL}}{1000 \text{ mL/L}} = 0.0100 \text{ L} = 1.00 \times 10^{-2} \text{ L}$

Concentration of $KOH_{(aq)} = 0.10$ mol/L

*Required:*

The pH of the mixture resulting from the reacting between 2.0 mL of 0.10 mol/L of $H_2SO_{4(aq)}$ and a 10.0 mL aliquot of 0.10 mol/L $KOH_{(aq)}$

*Strategy:*

This problem entails finding which of the two reagents is in excess. Knowing this, the final pH of the mixture can be found. First the number of moles of $H_2SO_{4(aq)}$ and $KOH_{(aq)}$ added; this would enable the amount of the excess reagent to be found. Knowing the number of moles of the excess reagent, its final concentration can be calculated since the final volume of the mixture is known; the sum of the volumes of the $H_2SO_{4(aq)}$ and $KOH_{(aq)}$ added together.

*Steps to be followed:*

**Step 1.** Find the number of moles of the $H_2SO_{4(aq)}$.

Using the equation, $\mathbf{n = C \times V}$, we get

$\mathbf{n} = 0.10$ mol/L $\times$ $2.0 \times 10^{-3}$ L $= 2.0 \times 10^{-4}$ L mol $H_2SO_{4(aq)}$

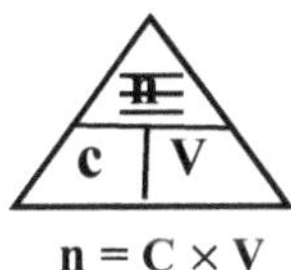


**Step 2.** Find the number of moles of the $KOH_{(aq)}$.

Using the equation, $\mathbf{n = C \times V}$, we get

$\mathbf{n} = 0.10$ mol/L $\times$ $1.00 \times 10^{-2}$ L $= \mathbf{1.00 \times 10^{-3}}$ **mol** $\mathbf{KOH_{(aq)}}$

**Step 3.** Find the excess reagent.

To do this first balance an equation to show the reaction between $KOH_{(aq)}$ and $H_2SO_{4(aq)}$, then choose one of the reagent to find how of the other one is required to react with it. The $H_2SO_{4(aq)}$ is chosen so becomes *the given reagent* and the $KOH_{(aq)}$ *required* one.

**Required**      **Given**

$2\ KOH_{(aq)}$ $+$ $H_2SO_{4(aq)}$ $\rightarrow$ $K_2SO_{4(aq)}$ $+$ $2\ H_2O_{(l)}$

**Step 5.** Find the number of moles of $KOH_{(aq)}$ required to react with $2.0 \times 10^{-4}$ L moles of $H_2SO_{4(aq)}$.

$$\frac{\text{Given}}{\text{Required}} = \frac{1}{2} = \frac{2.0 \times 10^{-4}\ \text{mol}\ H_2SO_{4(aq)}}{X\ \text{mol}\ KOH(aq)} = X = \mathbf{4.0 \times 10^{-4}}\ \textbf{L mol}\ \mathbf{KOH_{(aq)}}\ \textbf{(Required)}$$

Based on the calculations, it can be seen that $\mathbf{4.0 \times 10^{-4}}$ **L mol** $\mathbf{KOH_{(aq)}}$ *was required* to react with the $2.0 \times 10^{-4}$ L mol $H_2SO_{4(aq)}$, **but** $\mathbf{1.0 \times 10^{-3}}$ **mol** *was present.* This means that there was an excess of $KOH_{(aq)}$ in the mixture.

**Step 6.** Find the number of moles of $KOH_{(aq)}$ in excess.

$$\text{Excess}\ KOH_{(aq)} = 1.0 \times 10^{-3}\ \text{mol} - 4.0 \times 10^{-4}\ \text{mol} = \mathbf{6.0 \times 10^{-4}}\ \textbf{mol}\ \mathbf{KOH_{(aq)}}$$

**Step 7.** Find the concentration of the $KOH_{(aq)}$ in the mixture.

Final volume $= \mathbf{2.0 \times 10^{-3}}$ **L** (from $H_2SO_{4(aq)}$) $+ \mathbf{1.0 \times 10^{-2}}$ **L** (from $KOH_{(aq)}$) $= \mathbf{1.2 \times 10^{-2}}$ **L**

Using the formula, $\mathbf{C} = \dfrac{n}{V}$, we get $C = \dfrac{6.0 \times 10^{-4}\ \text{mol}}{1.2 \times 10^{-2}\ \text{L}}$

$$= \mathbf{5.0 \times 10^{-2}}\ \textbf{mol/L}\ \mathbf{KOH_{(aq)}}$$



**Step 8.** Find the concentration of the $[OH_{(aq)}]$ in the mixture. First show how $KOH_{(aq)}$ ionizes.

$KOH_{(aq)}$ $\rightarrow$ $K^+_{(aq)}$ $+$ $OH^-_{(aq)}$

$\mathbf{5.0 \times 10^{-2}}$ **mol/L**      $\mathbf{5.0 \times 10^{-2}}$ **mol/L**

**Step 9.** Find the $H^+_{(aq)}$ of the solution, using the formula.

Using the formula, $[OH^-_{(aq)}] \times [H^+_{(aq)}] = 1 \times 10^{-14}$ and making $[H^+_{(aq)}]$ the subject of the equation, we get

$$[H^+_{(aq)}] = \frac{1 \times 10^{-14}}{[OH^-]} = \frac{1 \times 10^{-14}\ \text{mol}}{5.0 \times 10^{-2}\ \text{L}} = \mathbf{2.0 \times 10^{-13}}\ \textbf{mol/L}$$

$$pH = -\log 2.0 \times 10^{-13} = \mathbf{12}$$

**Write the formulas for the following compounds.**

(a)  magnesium bromate $Mg(BrO_3)_2$

(b)  zinc iodite $Zn(IO_2)_2$

(c)  lead(II) sulphite  $PbSO_3$

(d)  ferric chlorate $Fe(ClO_3)_3$

(e)  tin(IV) carbonate  $Sn(CO_3)_2$

(f)  cupric acetate $Cu(CH_3COO)_2$

(g)  calcium hypoiodite $Ca(IO)_2$

(h) ammonium sulphate  $(NH_4)_2SO_4$

(i) potassium cyanide  $KCN$

(j) sodium periodate  $NaIO_4$

(k) plumbic nitrate  $Pb(NO_3)_4$

(l) potassium hydrogen sulphate  $KHSO_4$

(m) lithium phosphate  $Li_3PO_4$

(n) calcium nitrite  $Ca(NO_2)_2$



**Write the Classical or IUPAC names for the following compounds.**

(a)  $AgIO_2$ silver iodite

(b)  $MgCO_3$ magnesium carbonate

(c)  $Ca(HSO_4)_2$ calcium hydrogen sulphate

(d)  $Na_2SO_4.7H_2O$ sodium sulphate heptahydrate

(e)  $HBrO_{3(aq)}$ aqueous hydrogen bromate

(f)  $Sn(NO_3)_2$ tin(II) nitrate

(g)  $(NH_4)_2SO_4$ ammonium sulphate

(h)  $Pb(NO_2)_2$ lead(II) nitrite

(i)  $Mg(CH_3COO)_2$ magnesium acetate

(j) $K_3PO_4$ potassium phosphate

(k) $Be(CN)_2$ beryllium cyanide

(l) $Ba(HCO_3)_2$ barium hydrogen carbonate

(m) $SrSO_3$ strunctium sulphite

(n) $KClO$ potassium hypochlorite

(o) $KMnO_4$ potassium permanganate

(p) $HClO_{4(aq)}$ aqueous hydrogen perchlorate

(q) $Fe(IO_3)_3$ iron(III) iodate

(r) $Zn_3(PO_3)_2$ zinc phosphite

# Chapter Review: Acids and bases (17) solutions

## Matching

***Match each term in the table below with the correct statements that follow***

| | | | |
|---|---|---|---|
| A | $K_w = \left[H^+_{(aq)}\right] \times \left[OH^-_{(aq)}\right]$ | H | Aqueous ions |
| B | Strong acids | I | Weak acids and weak bases |
| C | Electrolytes | J | Brönsted-Lowry |
| D | Diprotic acids | K | Conjugate base |
| E | Titrant | L | Arrhenius theory |
| F | End point | M | pH scale |
| G | $H^+_{(aq)} + OH^-_{(aq)} \rightarrow H_2O_{(l)}$ | N | Standard solution |

| | |
|---|---|
| 1. | His theory states that acids produce $H^+_{(aq)}$ ions and bases produce $OH^-_{(aq)}$ ions respectively when they dissolve in water. |
| 2. | These species are responsible for the electrical conductivity in electrolytes. |
| 3. | Their theory states that an acid is a proton donor and a base is a proton acceptor. |
| 4. | These molecules ionize completely when they dissolve in water. |
| 5. | These molecules ionize only partially when they dissolve in water. |
| 6. | These conduct electricity when they dissolve in water. |
| 7. | This is what is left of the acid after it donates a proton to an acceptor. |
| 8. | These are capable of producing two moles of $H^+_{(aq)}$ ions per mole of molecules. |
| 9. | The net ionic equation for an acid base neutralization. |
| 10. | The reagent that is placed in the burette during titration. |
| 11. | The solution of known concentration used during titration. |
| 12. | The ionic product of water equation. |
| 13. | Tells if a substance is neutral, basic or acidic and to what degree it is either basic or acidic. |
| 14. | The point in titration when equal amount of $OH^-_{(aq)}$ is titrated with equal amount of $H^+_{(aq)}$ that causes the indicator to change colour. |

**True or False**

**Read each of the following statements and then decide if it is *True* or *False*.**

| 1. | Strong acids ionize in water to produce a low concentration of $H^+_{(aq)}$. |
|---|---|
| 2. | Weak acids have higher pH than stronger acids because they fully ionize when they dissolve in water. |
| 3. | A pH scale measures the $[H^+_{(aq)}]$ in a solution. |
| 4. | Unripe fruits taste sour because of the weak acids that they contain. |
| 5. | A diprotic acid produces only one mole of $H^+_{(aq)}$ ions. |
| 6. | Antacids are composed of strong bases. |
| 7. | A Brönsted-Lowry acid is a proton donor. |
| 8. | Bases have pH values lower than 7. |
| 9. | In an acid solution, $[H^+_{(aq)}] > [OH^-_{(aq)}]$. |
| 10. | A weak acid cannot neutralize a strong base. |
| 11. | Dilution of an acid solution raises its pH. |
| 12. | Pure water at temperatures above 25 $^0$C may have a pH greater than 7 |
| 13. | A dilute solution of a strong acid has lower pH than a concentrated solution of a weak acid. |
| 14. | If the sting of insects is caused by formic acid, the treatment for which is made of weak bases. |

## Multiple Choice

**Choose the letter that best answers the questions.**

1. A compound that ***dissociates*** in water and conducts electricity very well is most likely to be

| | | | | |
|---|---|---|---|---|
| a | a weak acid | d. | a soluble salt |
| b | a metal | e. | none of the above |
| c | a weak base | | |

2. Sodium chloride dissolves in water as follows: $NaCl_{(s)} + H_2O_{(l)} \rightarrow Na^+_{(aq)} + Cl^-_{(aq)}$.

   This change can be classified as

| | | | | |
|---|---|---|---|---|
| a | dissociation | d. | a and c only |
| b | melting | e. | none of the above |
| c | ionization | | |

3. Acetic acid (vinegar) is classified as a weak acid because

| | | | | |
|---|---|---|---|---|
| a | it is more than 99 % ionized | d. | it cannot neutralize strong bases |
| b | it ionizes about 50 % | e. | none of the above |
| c | it is only 1.3 % ionized | | |

**4.**

Carbonic acid ionizes as follows in water: $H_2CO_{3(aq)} + H_2O_{(l)} \rightleftharpoons HCO_3^-{}_{(aq)} + H_3O^+{}_{(aq)}$

The conjugate acid-base pairs produced are

| | | | |
|---|---|---|---|
| a | $H_2CO_3{}_{(aq)}$ and $H_2O_{(l)}$ | d | $H_2CO_{3(aq)}$ and $HCO_3^-{}_{(aq)}$ |
| b | $HCO_3^-{}_{(aq)}$ and $H_3O^+{}_{(aq)}$ | e | d and c only |
| c | $H_2O_{(l)}$ and $H_3O^+{}_{(aq)}$ | | |

**5.**

At 25 °C which of the following is true for pure water?

| | | | |
|---|---|---|---|
| a | $[OH^-{}_{(aq)}] = 1 \times 10^{-7}$ mol/L | d. | Its pH = 7.0 |
| b | $[H^+{}_{(aq)}] = 1 \times 10^{-7}$ mol/L | e. | all of the previous |
| c | $[H^+{}_{(aq)}] = [OH^-{}_{(aq)}]$ | | |

**6.**

If the pH of an aqueous solution is 9, then its pOH must be

| | | | |
|---|---|---|---|
| a | 9 | d | $1 \times 10^{-9}$ |
| b | 5 | e | $1 \times 10^{-5}$ |
| c | 14 | | |

**7.**

Which of the following is **not** true for a 0.10 mol/L $HCl_{(aq)}$ (a strong acid)?

| | | | |
|---|---|---|---|
| a | It is > 99% ionized | d. | It has a pH of 1.0 |
| b | It produces 0.1 mol/L $H^+{}_{(aq)}$ | e. | It has a pH of 6.0 |
| c | It is very sour | | |

**8.**

If an aqueous solution has a pOH of 11, its $[H^+{}_{(aq)}]$ must be

| | | | |
|---|---|---|---|
| a | $1.0 \times 10^{-3}$ mol/L | d | $3.0 \times 10^{-1}$ mol/L |
| b | $11.0 \times 10^{-3}$ mol/L | e | none of the above |
| c | $14.0 \times 10^{-3}$ mol/L | | |

**9.**

If the $[OH^-{}_{(aq)}]$ of an aqueous solution is $1.0 \times 10^{-9}$ mol/L, its $[H^+{}_{(aq)}]$ must be

| | | | |
|---|---|---|---|
| a | $1.0 \times 10^{-14}$ mol/L | c | $1.0 \times 10^{-7}$ mol/L |
| b | $1.0 \times 10^{-9}$ mol/L | d | none of the above |
| c | $1.0 \times 10^{-5}$ mol/L | | |

**10.**

Water is evaporated a basic solution of pH 10, which of the following will occur?

| | | | |
|---|---|---|---|
| a | $[OH^-{}_{(aq)}]$ will increase | d | pOH will decrease |
| b | $[H^+{}_{(aq)}]$ will decrease | e | all of the above |
| c | pH will increase | | |

11.

When $HI_{(aq)}$, a strong acid, is diluted with water, there would most likely be

| | | | |
|---|---|---|---|
| a | an increase in pH | d. | an increase in $[H^+_{(aq)}]$ |
| b | a decrease in pOH | e. | a, b, and c |
| c | a decrease in $[H^+_{(aq)}]$ | | |

12.

Hydrochloric acid of concentration $1.0 \times 10^{-3}$ mol/L will have a pH of

| | | | |
|---|---|---|---|
| a | 11.0 | d. | 1.0 |
| b | 3.0 | e | none of the above |
| c | 10 | | |

13.

Water reacts in the following two ways: $H_2O_{(l)} + H_2O_{(l)} \rightleftharpoons H_3O^+_{(l)} + OH^-_{(aq)}$

$$H_2O_{(l)} + HCl_{(aq)} \rightarrow H_3O^+_{(l)} + Cl^-_{(aq)}$$

Based on these reactions, water can be described as

| | | | |
|---|---|---|---|
| a | an acid | d | a, b and c |
| b | a base | e | None of the above |
| c | amphoteric | | |

14.

If acid A has a pH of 3 and another acid, B has a pH of 5, which of the following is true?

| | |
|---|---|
| a | Acid B is two times more acidic than acid A |
| b | Acid A is one hundred times more acidic than acid B |
| c | Acid A is one hundred times more concentrated with $[H^+_{(aq)}]$ than acid B |
| d | Acid B is two hundred times more concentrated with $[H^+_{(aq)}]$ than acid A |
| e | b and c |

15.

Sodium hydroxide reacts with chloric acid according to the equation below:

$$NaOH_{(aq)} + HClO_{3(aq)} \rightarrow NaClO_{3(aq)} + H_2O_{(l)}$$

Which of the following statements is **not** true?

| | |
|---|---|
| a | 25.0 mL of 0.1 mol/L $HClO_{3(aq)}$ will completely neutralize 50.0 mL of 0.05 mol/L $NaOH_{(aq)}$ |
| b | 100 mL of 0.05 mol/L $HClO_{3(aq)}$ will completely neutralize 50.0 mL of 0.1 mol/L $NaOH_{(aq)}$ |
| c | One mole of $NaOH_{(aq)}$ neutralizes exactly one mole of $HClO_{3(aq)}$ |
| d | A basic solution is produced if 50.0 mL of 0.20 mol/L $HClO_{3(aq)}$ is mixed with 40 mL of 0.1 mol/L $NaOH_{(aq)}$ |
| e | An acidic solution is produced if 25 mL of 0.2 mol/L $HClO_{3(aq)}$ is mixed with 15.0 mL of 0.30 mol/L $NaOH_{(aq)}$ |

16.

A 25 mL of 0.1 mol/L of $CH_3COOH_{(aq)}$ (**a weak acid**) will

| | |
|---|---|
| a | completely neutralize 25 mL of 0.10 mol/L $NaOH_{(aq)}$ (a strong base). |
| b | **not** completely neutralize 25 mL of 0.10 mol/L $NaOH_{(aq)}$. |
| c | completely neutralize 50 mL of 0.05 mol/L $Ca(OH)_{2(aq)}$ (a strong base). |
| d | **not** completely neutralize 50 mL of 0.05 mol/L $Ca(OH)_{2(aq)}$. |
| e | a and c |

17.

Which of the following is **not** true of antacids?

| | | | |
|---|---|---|---|
| a | They neutralize acids | d | They are weak bases |
| b | They have a pH greater than 7 | e | They are used to treat heart burns |
| c | They are strong bases | | |

18.

The bond in HF molecule is much more polar bond than HI molecule, yet $HI_{(aq)}$ is a very much stronger acid than $HF_{(aq)}$. The reason for this is due to the fact that

| | |
|---|---|
| a | the iodine atom is so much bigger than the fluorine atom that it allows the $HI_{(aq)}$ to ionize much easier than $HF_{(aq)}$. |
| b | iodine is more electronegative than fluorine |
| c | fluorine exerts a greater attractive force on hydrogen than iodine does |
| d | the polarity of bonds do not affect how easily a molecules ionize in water |
| e | a and c |

19.

Which of the following would not cause a change in pH, when added to water?

| | | | |
|---|---|---|---|
| a | $Mg(OH)_2$ | d | $C_2H_4O_2$ |
| b | NaCl | e | $NaH_2PO_4$ |
| c | HClO | | |

20.

Which of the following when dissolved in water would produce basic solutions?

| | | | |
|---|---|---|---|
| a | $Na_2O_{(s)}$ | d | $NH_{3(g)}$ |
| b | $SO_{2(g)}$ | e | a and d |
| c | $NaCl_{(s)}$ | | |

21.

Which of the following is true of $NaHSO_4$?

| | |
|---|---|
| a | It is a basic salt |
| b | It is an acidic salt |
| c | one mole of it will completely neutralize one mole of NaOH |
| d | It is a neutral salt |
| e | a and c |

**Matching:**

| | | | | | | | | |
|---|---|---|---|---|---|---|---|---|
| 1.) L | 2.) H | 3.) J | 4.) B | 5.) I | 6.) C | 7.) K | 8.) D | 9.) G |
| 10.) E | 11.) N | 12.) A | 13.) M | 14.) F | | | | |

**True/False:**

| | | | | | | | | |
|---|---|---|---|---|---|---|---|---|
| 1.) F | 2.) F | 3.) T | 4.) T | 5.) F | 6.) F | 7.) T | 8.) F | 9.) T |
| 10.) F | 11.) T | 12.) T | 13.) T | 14.) T | | | | |

**Multiple Choice:**

| | | | | | | | | |
|---|---|---|---|---|---|---|---|---|
| 1.) D | 2.) A | 3.) C | 4.) E | 5.) E | 6.) B | 7.) E | 8.) A | 9.) E |
| 10.) E | 11.) E | 12.) B | 13.) D | 14.) E | 15.) D | 16.) E | 17.) C | 18.) A |
| 19.) B | 20.) E | 21.) E | | | | | | |

# Unit 6: Gases



**Boyle's Law problems**

*Problem:*

1. A balloon filled with an inert gas has an initial volume of 2.0 L and pressure of 120.0 kPa.
    a) What will happen to its volume if it ascends into the atmosphere at constant temperature?
    b) What will be its final pressure if its new volume was 3.0 L?

*Provided:*

Initial volume 2.0 L ($V_1$)

Initial pressure 120.0 kPa ($P_1$)

*Required:*

What will happen if the balloon ascends into the atmosphere at constant temperature?

What will be the final pressure ($P_2$) the balloon if its new volume was 3.0 L ($V_2$)?

*Solution:*

a) Because the pressure of the atmosphere gets progressively less upwards, the walls of the balloon experience this and expand proportionately.
b) Using the equation, $P_1 \times V_1 = P_2 \times V_2$ and making $P_2$ and then substituting values, we get

$$P_2 = \frac{P_1 \times V_1}{V_2} = \frac{120.0 \text{ kPa} \times 2.0 \text{ L}}{3.0 \text{ L}} = \textbf{80.0 kPa}$$

*Problem:*

2. A cylinder containing 5.0 L helium gas is at 4.0 atm, is used to fill balloons to a volume of 500.0 mL at a pressure of 1.0 atm. How many balloons can be filled?

*Provided:*

Initial volume of balloon = 5.0 L ($V_1$)

Initial pressure of balloon = 4.0 atm ($P_1$)

The volume of each balloon = 500.0 mL = $\dfrac{500.0 \text{ mL}}{1000 \text{ mL/L}}$ = 0.5000 L

The pressure of each balloon = 1.0 atm

*Required:*

The number of balloons of volume 500.0 mL and pressure 1.0 atm each that can be filled from a cylinder containing 5.0 L helium gas is at 4.0 atm.

*Strategy:*

Find the total volume of helium that can be produced at 1.0 atm, and then divide that by the volume of one balloon, assuming that all the helium from the cylinder comes out. If not, subtract the volume of the helium that remains in the cylinder, (5.0 L) from the total volume and the then divide by the volume of one balloon.

*Solution:*

**Step 1.** Using the equation, $P_1 \times V_1 = P_2 \times V_2$ and making $V_2$ and then substituting values, we get

$$V_2 = \frac{P_1 \times v_1}{P_2} = \frac{4.0 \text{ atm} \times 5.0 \text{ L}}{1.0 \text{ atm}} = 20.0 \text{ L}$$

**Step 2.** Divide the total volume by 0.500 L

$$= \frac{20.0 \text{ L}}{0.500 \text{ L/balloon}} = \textbf{40 balloons}$$

If 5.0 L of helium remains, then the number of balloons that can be filled will be:

$$(20.0 \text{ L} - 5.0 \text{L}) \div 0.500 \text{ L/balloon} = \textbf{30 balloons}$$

*Problem:*

3. A cylinder contains air of volume of 2.0 L at pressure 1.0 atm. If the walls of the cylinder can only withstand a maximum pressure of 4.0 atm, to what critical volume can the air in it be compressed before the cylinder bursts?

*Provided:*

Initial volume = 2.0 L ($V_1$)

Initial pressure = 1.0 atm ($P_1$)

Final pressure = 4.0 atm ($P_2$)

*Required:*

Final volume ($V_2$)

*Solution:*

Using the equation, $P_1 \times V_1 = P_2 \times V_2$ and making $V_2$ and then substituting values, we get

$$V_2 = \frac{P_1 \times v_1}{P_2} = \frac{1.0 \text{ atm} \times 2.0 \text{ L}}{4.0 \text{ atm}} = \textbf{0.50 L}$$

*Problem:*

4. A syringe has 40.0 mL of air at 101.0 kPa. What pressure must be applied to the gas for it to have a final volume of 30.0 mL, assuming that the mass and temperature of the gas stay constant.

*Provided:*

Initial volume = 40.0 mL ($V_1$)

Initial pressure = 101.0 kPa ($P_1$)

Final volume = 30.0 mL ($V_2$)

*Required:*

Final pressure ($P_2$)

*Solution:*

Using the equation, $P_1 \times V_1 = P_2 \times V_2$ and making $P_2$ and then substituting values, we get

$$P_2 = \frac{P_1 \times v_1}{V_2} = \frac{101.0 \text{ kPa} \times 40.0 \text{ mL}}{30.0 \text{ mL}} = \textbf{135 kPa}$$

1. Convert the following Celsius temperature to Kelvin:

   To do these, add 273 to existing values and then change the unit to Kelvin.

   a) $10\,^{\circ}C$ $\longrightarrow$ 283 K

   b) $100\,^{\circ}C$ $\longrightarrow$ 373 K

   c) $120\,^{\circ}C$ $\longrightarrow$ 393 K

   d) $-50\,^{\circ}C$ $\longrightarrow$ 223 K

2. Convert the following Kelvin values to Celsius temperature:

   To do these, subtract 273 to existing values and then change the unit to Celcius.

   a) 75 K $\longrightarrow$ $-198\,^{\circ}C$

   b) 120 K $\longrightarrow$ $-153\,^{\circ}C$

   c) 373 K $\longrightarrow$ $100\,^{\circ}C$



## Charles' Law problems

*Problem*:

**a)** Air in a cylinder with a movable piston has a volume of 4.00 L at 20.0 °C. To what temperature must the air be warmed to reach a new volume of 4.50 L at constant pressure?

*Provided:*

   Initial volume $V_1$ = 4.0 L

   Initial temperature = $T_1$ = 20.0 °C = 293 K

   Final volume = 4.50 L ($V_2$)

*Required:*

   Final temperature = ($T_2$)

## Solution:

Using the formula, $\dfrac{V_1}{T_1} = \dfrac{V_2}{T_2}$ and making $T_2$ the subject of the equation and the substituting the values, we get

$$T_2 = \frac{V_2 \times T_1}{V_1} = \frac{4.50\ L \times 293\ K}{4.0\ L} = 330\ K = \mathbf{57\ ^{\circ}C}$$

*Problem*:

**b)** Inside a house, a balloon containing helium gas at 22.0 °C has a volume of 5.0 L. What will be its volume if it is placed outside where temperature is lowered to 5 °C? Assume that that the atmospheric pressure remains constant.

*Provided:*

Initial volume $V_1$ = 5.0 L

Initial temperature $T_1$ = 22.0 °C = 295 K

Final temperature $T_2$ = 5 °C = 278 K

*Required:*

Final volume: $V_2$

**Solution:**

Using the formula, $\dfrac{V_1}{T_1} = \dfrac{V_2}{T_2}$ and making $V_2$ the subject of the equation and the substituting the values, we get

$$V_2 = \frac{V_1 \times T_2}{T_1} = \frac{5.0 \text{ L} \times 278 \text{ K}}{295 \text{ K}} = \mathbf{4.7 \text{ L}}$$

**Problem:**

c)  2.0 L of argon gas is heated at 20.0 °C until its final temperature is 32.0 °C at constant pressure. What is its final volume?

*Provided:*

Initial volume $V_1$ = 2.0 L

Initial temperature $T_1$ = 20.0 °C = 293 K

Final temperature $T_2$ = 32.0 °C = 305 K

*Required:*

Final volume: ($V_2$)

**Solution:**

Using the formula, $\dfrac{V_1}{T_1} = \dfrac{V_2}{T_2}$ and making $V_2$ the subject of the equation and the substituting the values we get

$$V_2 = \frac{V_1 \times T_2}{T_1} = \frac{2.0 \text{ L} \times 305 \text{ K}}{293 \text{ K}} = \mathbf{2.1 \text{ L}}$$

**Problem:**
d)  A balloon filled with only helium gas had a volume of 4.00 L at 22.0 °C. °C. When placed outdoors for one cold night it attained a new volume of 3.75 L. Assuming that the atmospheric pressure remained constant, what was the highest temperature that balloon was subjected to?

*Provided:*

Initial volume: 4.0 L ($V_1$)

Initial temperature: $T_1$ = 22.0 °C = 295 K

Final volume: 3.75 L ($V_2$)

*Required:*

Final temperature: ($T_2$)

*Solution:*

Using the formula, $\dfrac{V_1}{T_1} = \dfrac{V_2}{T_2}$ and making $T_2$ the subject of the equation and the substituting the values, we get

$$T_2 = \frac{V_2 \times T_1}{V_1} = \frac{3.75 \text{ L} \times 295 \text{ K}}{4.00 \text{ L}} = 277 \text{ K} = \mathbf{3.56 \ ^\circ C}$$

## Exercise 18.4 (p.333)

**Pressure and Temperature Law problems**

*Problem:*

**a)** A sealed container has 2.00 L of methane gas at a pressure of 120.0 kPa and temperature of 20.0 °C. If the container is placed in a room where the temperature is 35.0 °C, what will be its new pressure?

*Provided:*

      Initial pressure ($P_1$) = 120.0 kPa

      Initial temperature ($T_1$) = 20.0 °C = 293 K

      Final temperature ($T_2$) = 35.0 °C = 308 K

*Required:*

      Final pressure: ($P_2$)

*Solution:*

      In this problem, the volume remains constant. All temperatures must be in Kelvin.

Using the formula, $\dfrac{P_1}{T_1} = \dfrac{P_2}{T_2}$ and making $P_2$ the subject of the equation and the substituting the values, we get

$$P_2 = \frac{T_2 \times P_1}{T_1} = \frac{308 \text{ K} \times 120.0 \text{ kPa}}{293 \text{ K}} = \mathbf{126 \ kPa}$$

*Problem:*

**b)** A cylinder of hydrogen gas at 22.0 °C has a pressure of 4.00 atm. If the maximum internal pressure that the cylinder can withstand is 12.00 atm, find the minimum temperature at which the cylinder would explode if it were in a building where a fire broke out?

*Provided:*

      Initial temperature:($T_1$) = 22.0 °C = 295 K

      Initial pressure: ($P_1$) = 4.00 atm

      Final pressure: ($P_2$) = 12.00 atm

*Required:*

      Final temperature: ($T_2$)

*Solution:*

In this problem, the volume remains constant. All temperatures must be in Kelvin.

Using the formula, $\dfrac{P_1}{T_1} = \dfrac{P_2}{T_2}$ and making $T_2$ the subject of the equation and the substituting the values, we get

$$T_2 = \frac{P_2 \times T_1}{P_1} = \frac{12.00\ \text{kPa} \times 295\ \text{K}}{4.00\ \text{kPa}} = 885\ \text{K} = \mathbf{612\ ^{\circ}C}$$

*Problem:*

**c)** The volume of gas above a half-filled soda glass bottle is 500.0 mL at a temperature of 25.0 °C and pressure of 110.0 kPa. What will be the new pressure of the gas if the bottle is placed in a refrigerator where it is cooled to 5.0 °C and 10% of the gas particles dissolved in the soda?

*Provided:*

Initial pressure: $(P_1)$ = 110.0 kPa

Initial temperature: $(T_1)$ = 25.0 °C = 298 K

Final temperature: $(T_2)$ = 5.0 °C = 278 K

*Required:*

Final pressure: $(P_2)$

*Solution:*

In this problem, the volume remains constant. All temperatures must be in Kelvin.

Using the formula, $\dfrac{P_1}{T_1} = \dfrac{P_2}{T_2}$ and making $P_2$ the subject of the equation and the substituting the values, we get

$$P_2 = \frac{T_2 \times P_1}{T_1} = \frac{278\ \text{K} \times 110.0\ \text{kPa}}{298\ \text{K}} = \mathbf{102.6\ kPa}$$

Since 10% of the gas has dissolved into the soda because of higher solubility at a lower temperature, the pressure of the gas above the liquid will decrease proportionately; it would be only 90 % of the total.

$$\text{The real pressure} = \frac{102.6\ \text{kPa} \times 90\ \%}{100\ \%} = \mathbf{92.34\ kPa}$$

*Problem:*

**d)** Assuming that after several hours of continuous driving at high speed the maximum temperature that tires reach is 50.0 °C. What must be the safe initial pressure that tires at 20.0 °C must be set, before going on a long journey if the maximum pressure that the tires can withstand is 280.0 kPa? Assume that the volume of the tires stays constant.

*Provided:*

Initial temperature $(T_1)$ = 20.0 °C = 293 K

Final pressure $(P_2)$ = 280.0 kPa

Final temperature $(T_2)$ = 50.0 °C = 328 K

*Required:*

Final pressure: $(P_1)$

*Solution*:

In this problem, the volume remains constant. All temperatures must be in Kelvin.

Using the formula, $\dfrac{P_1}{T_1} = \dfrac{P_2}{T_2}$ and making $P_1$ the subject of the equation and the substituting the values, we get

$$P_1 = \frac{P_2 \times T_1}{T_2} = \frac{280.0\ \text{kPa} \times 293\ \text{K}}{328\ \text{K}} = \textbf{250.1 kPa}$$

### Exercise 18.5 (p.336-337)

## The Combined Gas Law problems

### *Problem:*

**a)** Helium gas is stored in a 10.0 L steel cylinder at 20.0 °C and 1200.0 kPa pressure. How many balloons of capacity of 2.0 L can be filled with helium gas at 25.0 °C and 120.0 kPa from the cylinder?

*Provided:*

Initial temperature: $(T_1)$ = 20.0 °C = 293 K

Initial volume = 10.0 L

Initial pressure: $(P_2)$ = 1200.0 kPa

Final pressure: $(P_2)$ = 120.0 kPa

Final temperature: $(T_2)$ = 25.0 °C = 298 K

*Required:*

The number of balloons of capacity of 2.0 L can be filled with helium gas at 25.0 °C and 120.0 kPa from the cylinder?

*Strategy:*

To solve this problem, find the volume of helium gas at the new conditions; 120.0 kPa and 298 K and then divide it by the volume of one balloon; 2.0 L, assuming that all the helium exits the cylinder. If not, subtract the volume of the cylinder from the calculated volume then divide by the volume of one balloon.

*Steps to be followed:*

**Step 1.** Using the combined gas equation, $\dfrac{P_1 \times V_1}{T_1} = \dfrac{P_2 \times V_2}{T_2}$ and making $V_2$ the subject of equation, we get

$$V_2 = \frac{P_1 \times V_1 \times T_2}{T_1 \times P_2}$$ . Substituting values into this equation we get,

$$V_2 = \frac{1200.0\ \text{kPa} \times 10.0\ \text{L} \times 298\ \text{K}}{293\ \text{K} \times 120.0\ \text{kPa}} = \textbf{102 L}\ \text{of helium gas.}$$

**Step 2.** Find the number of balloons in both scenarios:

$$= \frac{102\ \text{L}}{2.0\ \text{L/balloon}} = \textbf{51 balloons}\ \text{or} = \frac{92\ \text{L}}{2.0\ \text{L/balloon}} = \textbf{46 balloons}$$

## Problem:

**a)** A weather balloon contains 6.00 L of an inert gas at 15.0 °C and 90.0 kPa pressure, somewhere in the atmosphere. If it is brought to the surface of the Earth where the new temperature and pressure are 25.0 °C and 101.0 kPa respectively, what would be its new volume.

*Provided:*

Initial temperature $(T_1)$ = 15.0 °C = 288 K

Initial volume $(V_1)$ = 6.00 L

Initial pressure $(P_2)$ = 90.0 kPa

Final pressure $(P_2)$ = 101.0 kPa

Final temperature $(T_2)$ = 25.0 °C = 298 K

*Required:*

Final volume $(V_2)$

*Steps to be followed:*

**Step 1.** Using the combined gas equation, $\dfrac{P_1 \times V_1}{T_1} = \dfrac{P_2 \times V_2}{T_2}$ and making $V_2$ the subject of equation, we get

$V_2 = \dfrac{P_1 \times V_1 \times T_2}{T_1 \times P_2}$ . Substituting values into this equation we get

$$V_2 = \frac{90.0 \text{ kPa} \times 6.0 \text{ L} \times 298 \text{ K}}{288 \text{ K} \times 101.0 \text{ kPa}} = \mathbf{5.5 \text{ L}}$$

## Problem:

**c)** A car tire filled with air has a pressure of 240.0 kPa, a volume of 12.0 L and temperature of 15.0 °C. If after several hours of high-speed driving, its temperature rose to 40.0 °C and its volume increased by 5.00%, what is the new pressure?

*Provided:*

Initial temperature $(T_1)$ = 15.0 °C = 288 K

Initial volume $(V_1)$ = 12.0 L

Initial pressure $(P_1)$ = 240.0 kPa

Final temperature $(T_2)$ = 40.0 °C = 313 K

*Required:*

Final pressure in the tires

*Strategy:*

To solve this problem, first find the new volume of the tires. Then make the final pressure $(P_2)$ the subject of the equation, followed by insertion of the various values into the equation.

*Steps to be followed:*

**Step 1.** Find the new volume $(V_2)$ of the tires.

$$V_2 = 12.0 \text{ L} + \frac{12.0 \text{ L}}{100 \text{ \%}} \times 5.00 \text{ \%} = 12.6 \text{ L}$$

**Step 2.** Using the combined gas equation, $\dfrac{P_1 \times V_1}{T_1} = \dfrac{P_2 \times V_2}{T_2}$ and making $P_2$ the subject of equation, we get

$P_2 = \dfrac{P_1 \times V_1 \times T_2}{T_1 \times V_2}$ . Substituting values into this equation we get

$$P_2 = \frac{240.0\ \text{kPa} \times 12.0\ \text{L} \times 313\ \text{K}}{288\ \text{K} \times 12.6\ \text{L}} = \mathbf{248\ kPa}$$

## *Problem:*

**d)** Oxygen gas collected above water at 25 °C and 102.5 kPa has a volume of 200.0 mL. What would be the volume of dry oxygen at STP? (The vapour pressure of water at 25.0 °C is 3.17 kPa).

*Provided:*

Initial temperature ($T_1$) = 25.0 °C = 298 K

Initial volume ($V_1$) = 200.0 mL = 0.200 L

Initial pressure: ($P_1$) = 102.5 kPa (water vapour + oxygen)

Final temperature ($T_2$) = 0 °C = 273 K

Final pressure ($P_2$) = 101.325 kPa

The vapour pressure of water at 25.0 °C = 3.17 kPa

*Required:*

Final volume of the dry oxygen ($V_2$)

*Strategy:*

Before the final volume of the dry oxygen can be calculated, its initial pressure volume must be found. This is done by subtracting the vapour pressure of water at 25.0 °C from the initial pressure.

*Steps to be followed:*

**Step 1.** Find the partial pressure of the **dry oxygen**

$P_1$ = 102.5 kPa - 3.17 kPa = **99.3 kPa**

**Step 2.** Using the combined gas equation, $\dfrac{P_1 \times V_1}{T_1} = \dfrac{P_2 \times V_2}{T_2}$ and making $V_2$ the subject of equation, we get

$V_2 = \dfrac{P_1 \times V_1 \times T_2}{T_1 \times P_2}$ . Substituting values into this equation we get

$$V_2 = \frac{99.3\ \text{kPa} \times 0.200\ \text{L} \times 2\,73\ \text{K}}{298\ \text{K} \times 102.5\ \text{kPa}} = \mathbf{0.176\ L = 176\ mL}$$

## *Problem:*

**e)** At what new temperature must 10.0 L of $CO_{2(g)}$ at 20.0 °C and 100.2 kPa be heated to have a volume of 12.5 L at 100.0 kPa?

*Provided:*

Initial temperature ($T_1$) = 20.0 °C = 293 K

Initial volume ($V_1$) = 10.0 L

Initial pressure ($P_1$) = 100.2 kPa

Final pressure ($P_2$) = 100.0 kPa

Final volume ($V_2$) = 12.5 L

Final temperature (T$_2$)

*Steps to be followed:*

**Step 1.** Using the combined gas equation, $\dfrac{P_1 \times V_1}{T_1} = \dfrac{P_2 \times V_2}{T_2}$ and making T$_2$ the subject of equation, we get

$T_2 = \dfrac{P_2 \times V_2 \times T_1}{P_1 \times V_1}$ . Substituting values into this equation we get

$$\mathbf{T_2} = \dfrac{100.0 \text{ kPa} \times 12.5 \text{ L} \times 293 \text{ K}}{100.2 \text{ kPa} \times 10.0 \text{ L}} = 365.5 \text{ K} = \mathbf{92.5\ ^oC}$$

**Problem:**

**f)** 24.0 L of carbon dioxide gas is at a temperature of 25 $^\circ$C and at a pressure of 100.0 kPa. To what temperature must the gas be cooled to attain a new volume of 14.0 L at 98.0 kPa?

*Provided:*

Initial temperature (T$_1$) = 25.0 $^\circ$C = 298 K

Initial volume (V$_1$) = 24.0 L

Initial pressure (P$_1$) = 100.0 kPa

Final pressure (P$_2$) = 98.0 kPa

Final volume (V$_2$) = 14.0 L

*Required:*

Final temperature: (T$_2$)

*Steps to be followed:*

**Step 1.** Using the combined gas equation, $\dfrac{P_1 \times V_1}{T_1} = \dfrac{P_2 \times V_2}{T_2}$ and making T$_2$ the subject of equation, we get

$T_2 = \dfrac{P_2 \times V_2 \times T_1}{P_1 \times V_1}$ . Substituting values into this equation we get

$$\mathbf{V_2} = \dfrac{98.0 \text{ kPa} \times 14.0 \text{ L} \times 298 \text{ K}}{100.0 \text{ kPa} \times 24.0 \text{ L}} = \mathbf{170 \text{ K}} = \mathbf{- 103\ ^oC}$$

**Problem:**

**g)** To what new pressure 10.4 L of nitrogen at 101.2 kPa and 25 $^\circ$C be raised, if its temperature is raised to 60 $^\circ$C?

*Provided:*

Initial temperature (T$_1$) = 25.0 $^\circ$C = 298 K

Final temperature (T$_2$) = 60 $^\circ$C = 333 K

Initial volume (V$_1$) = 10.4 L

Initial pressure (P$_1$) = 101.2 kPa

*Required:*

Final pressure (P$_2$)

*Solution:*

In this problem it is assumed that the volume of the gas remains constant.

*Steps to be followed:*

**Step 1.** Using the formula, $\dfrac{P_1}{T_1} = \dfrac{P_2}{T_2}$ and making $P_2$ the subject of the equation and the substituting the values, we get

$$P_2 = \frac{P_1 \times T_2}{T_1} = \frac{101.2 \text{ kPa} \times 333 \text{ K}}{298 \text{ K}} = 113 \text{ kPa}$$

## Exercise 19.1 (p.340)

1)  1 atm.          2)

| Gases | Inhaled air | Exhaled air |
|---|---|---|
| $O_2$ | > pp | < pp |
| $CO_2$ | < pp | > pp |
| $H_2O$ | < pp | > pp |

## Exercise 19.2 (p.346-345)

**Molar Volume problems**

*Problem:*

**1.** Determine the number of molecules in 33.6 L of $N_{2(g)}$ at STP.

*Provided:*

Volume of $N_{2(g)}$ at STP = 33.6 L

*Required:*

The number of molecules of nitrogen in 33.6 L at STP.

*Solution:*

- *Recall that one mole of a gas at STP has a volume of 22.4 L.*

**Step 1.** Find the number of moles in 33.6 L of $N_{2(g)}$

$$\frac{33.6 \text{ L}}{22.4 \text{ L/mol}} = 1.50 \text{ mol}$$

**Step 2.** Find the number of molecules (N) in 1.50 mol of $N_{2(g)}$

Using the formula, $N = n \times N_A$ we get, $N = 1.50 \text{ mol} \times 6.02 \times 10^{23}$ molecules/ mol

$$= 9.03 \times 10^{23} \text{ molecules } 10^{23}$$

*Problem:*

**2.** Find the volume of $O_{2(g)}$ at SATP that is produced from the decomposition of 3.40 g of $H_2O_{2(l)}$. The following equation shows how this decomposition happens:

$$\underset{\text{Given}}{2\ H_2O_{2(l)}} \quad \rightarrow \quad 2\ H_2O_{(l)} \quad + \quad \underset{\text{Required}}{O_{2(g)}}$$

*Provided:*

Mass of $H_2O_{2(l)}$ = 3.40 g

Molar volume at SATP = 24.8 L

*Required:*

The volume of $O_{2(g)}$ at SATP that is produced from the decomposition of 3.40 g of $H_2O_{2(l)}$.

*Strategy:*

First find the number of moles of $H_2O_2$ in 3.40 g, then use the mole ratios in the equation to find the number of moles of $O_{2(g)}$ produced. Knowing the number of moles of $O_{2(g)}$ produced, its volume can be calculated.

**Step 1.** Find the number of moles of $H_2O_2$ in 3.40 g. First find the molar mass of $H_2O_2$.

$$M = 2 \times 1.01 \text{ g/mol H} + 2 \times 16.00 \text{ g/mol O} = \textit{34.02 g/mol}$$

Using the formula, $\mathbf{n = \dfrac{m}{M}}$ we get $n = \dfrac{3.40 \text{ g}}{34.02 \text{ g/mol}} = \mathbf{0.100 \text{ mol } H_2O_2}$

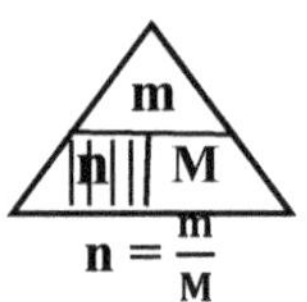

**Step 2.** Find the number of moles of $O_{2(g)}$ produced

$$\frac{\text{Given}}{\text{Required}} \ \frac{2}{1} = \frac{0.100 \text{ mol } H_2O_2}{X \text{ mol } O_2} = 2 \ X = 0.100 \text{ mol } O_{2(g)}$$

$$\mathbf{X = 0.0500 \text{ mol } O_{2(g)}}$$

**Step 3.** Find the volume of $O_{2(g)}$ produced.

$$\mathbf{V} = 0.0500 \text{ mol} \times 24.8 \text{ L/mol} = \mathbf{1.24 \text{ L}}$$

## Problem:

3. The following equation shows how methane gas is burnt in oxygen:

$$\underset{\text{Given}}{CH_{4(g)}} + \underset{\text{Required}}{2 \ O_{2(g)}} \rightarrow 2H_2O_{(l)} + CO_{2(g)}$$

What volume of oxygen at SATP is required for complete combustion of 4.00 g of methane?

*Provided:*

Mass of methane = 4.00 g

Molar volume at SATP = 24.8 L

*Required:*

Volume of oxygen at SATP that is required for complete combustion of 4.00 g of methane.

*Strategy:*

First find the number of moles of $CH_{4(g)}$ in 4.00 g, then use the mole ratios in the equation to find the number of moles of $O_{2(g)}$ required to react with it. Knowing the number of moles of $O_{2(g)}$ produced, its volume can be calculated.

**Step 1.** Find the number of moles of $CH_{4(g)}$ in 4.00 g. First find the molar mass of $CH_4$.

$$M = 1 \times 12.01 \text{ g/mol C} + 4 \times 1.01 \text{ g/mol H} = \mathbf{16.05 \text{ g/mol}}$$

Using the formula, $\mathbf{n = \dfrac{m}{M}}$ we get $\mathbf{n} = \dfrac{4.00 \text{ g}}{16.05 \text{ g/mol}} = \mathbf{0.250 \text{ mol } CH_4}$

**Step 2.** Find the number of moles of $O_{2(g)}$ produced.

$$\frac{\text{Given}}{\text{Required}} \quad \frac{1}{2} = \frac{0.250 \text{ mol CH}_4}{\text{X mol O}_2} = X = 0.500 \text{ mol O}_{2(g}$$

**Step 3.** Find the volume of $O_{2(g)}$ produced.

$$V = 0.500 \text{ mol} \times 24.8 \text{ L/mol} = \textbf{12.4 L } O_{2(g)}$$

## *Problem:*

**4.** Ammonia ($NH_{3(g)}$) is synthesised by the reaction between nitrogen and hydrogen. What volume of nitrogen and hydrogen will be produced at SATP by the decomposition of 74.49 L of $NH_{3(g)}$ at SATP?

*Provided:*

Volume of $NH_{3(g)}$ at SATP = 74.49 L

Molar volume at SATP = 24.8 L

*Required:*

Volumes of nitrogen and hydrogen that will be produced at SATP by the decomposition of

*Strategy:*

First find the number of moles of $NH_{3(g)}$ in 74.49 L, then use the mole ratios in the equation to find the number of moles of nitrogen and hydrogen that will be produced. Knowing their number of moles, their volumes can be calculated.

**Step 1.** Write a balanced equation for the decomposition of $NH_{3(g)}$

$$\overset{\textbf{Given}}{2 \text{ NH}_{3(g)}} \quad \rightarrow \quad \overset{\textbf{Required}}{N_{2(g)}} \quad + \quad \overset{\textbf{Required}}{3 \text{ H}_{2(g)}}$$

**Step 2.** Find the volume of $NH_{3(g)}$ decomposed.

$$n = \frac{74.49 \text{ L}}{24.8 \text{ L/mol}} = \textbf{3.00 mol NH}_{3(g)}$$

**Step 3.** Find the number of moles of $N_{2(g)}$ produced

$$\frac{\text{Given}}{\text{Required}} \quad \frac{2}{1} = \frac{3.00 \text{ mol NH}_{3(g)}}{\text{X mol } N_{2(g)}} = 2 X = 0.100 \text{ mol } O_{2(g)}$$

$$X = \textbf{1.50 mol N}_{2(g)}$$

**Step 4.** Find the volume of 1.50 mol $N_{2(g)}$ at SATP.

$$V = 1.50 \times 24.8 \text{ L/mol} = \textbf{37.2 L N}_{2(g)}$$

**Step 5.** Find the number of moles of $H_{2(g)}$ produced.

$$\frac{\text{Given}}{\text{Required}} \quad \frac{2}{3} = \frac{3.00 \text{ mol NH}_{3(g)}}{\text{X mol } H_{2(g)}} = 2 X = 9.00 \text{ mol } H_{2(g)}$$

$$X = \textbf{4.50 mol H}_{2(g)}$$

**Step 6.** Find the volume of 4.50 mol $H_{2(g)}$ at SATP.

$$V = 4.50 \times 24.8 \text{ L/mol} = \textbf{112 L H}_{2(g)}$$

## *Problem:*

**5.** A 4-L cylinder contains 5.00 moles of compressed $He_{(g)}$. How many 2-L balloons at SATP can be filled by the compressed $He_{(g)}$, assuming all the gas is emptied?

*Provided:*

Volume of compressed = 4.00 L

Number of moles of compressed $He_{(g)}$ = 5.00 mol

Volume of each balloon at SATP = 2.0 L

*Required:*

The number of 2-L balloons at SATP can be filled by the compressed $He_{(g)}$, assuming all the gas is emptied.

*Strategy:*

Find the volume of 5.00 moles of the $He_{(g)}$ at SATP, then divide it by the volume of 1 balloon (2.0 L)

**Step 1.** Find the volume of 5.00 moles of the $He_{(g)}$ at SATP.

$$V = 5.00 \text{ mol} \times 24.8 \text{ L/mol} = \textbf{124 L } \textbf{He}_{(g)}$$

**Step 2.** Find the number of 2-L balloons that can be filled by 124 L of $He_{(g)}$.

$$= \frac{124 \text{ L}}{2 \text{ L/balloon}} = \textbf{62 balloons}$$

## Problem:

6.  Potassium chlorate decomposes according to the following equation:

**Required**             **Given**

$2 \text{ KClO}_{3(s)} \quad \rightarrow \quad 2 \text{ KCl}_{(s)} \; + \quad 3 \text{ O}_{2(g)}$

If the only source of oxygen for the burning of carbon is from the decomposition of potassium chlorate, what mass of it must decompose to produce 124.0 L of oxygen at SATP? What mass of carbon can be completely burnt by this volume of oxygen produced?

*Given:*

The volume of oxygen: 124.0 L at SATP

*Required:*

- Mass of potassium chlorate that must be decomposed to produce 124.0 L of oxygen at SATP
- The mass of carbon that can be completely burnt by 124.0 L of oxygen

*Strategy:*

To solve this problem, convert the volume of $O_{2(g)}$ to amount in moles, then use the mole ratios in the balanced equation to find the number of moles of $KClO_{3(s)}$ that must be decomposed. Knowing the number of moles of $KClO_{3(s)}$, its mass can be calculated.

To know the mass of carbon that can be burnt from the oxygen produced, first balance an equation for the reaction between carbon and oxygen. Knowing the numbers of moles of oxygen, the requisite number of moles of carbon can be calculated, from which its mass can be found.

**Step 1.** Find the number of moles of oxygen at SATP.

$$n = \frac{124.0 \text{ L}}{24.8 \text{ L/mol}} = \textbf{5.00 mol } \textbf{O}_{2(g)}$$

**Step 2.** Find the number of moles of $KClO_{3(s)}$, using the mole ratio in the balanced equation

$$\frac{\text{Given}}{\text{Required}} \; \frac{3}{2} = \frac{5.00 \text{ mol O}_{2(g)}}{\text{X mol KClO}_{3(s)}} = 3 \text{ X} = \textbf{10.0 mol KClO}_{3(s)}$$

$$\text{X} = \textbf{3.33 mol KClO}_{3(s)}$$

**Step 3.** Find the mass of 3.33 moles of $KClO_{3(s)}$. First find its molar mass.

$M = 1 \times 39.10$ g/mol K $+ 1 \times 35.45$ g/mol Cl $+ 3 \times 16.00$ g/mol O $= \textit{122.6 g/mol}$

Using the formula, $\mathbf{m = n \times M}$, we get

Mass $= 3.330$ mol $\times 122.6$ g/mol $= \mathbf{408.1\ g\ KClO_{3(s)}}$

**Step 4.** Balance an equation for the reaction between carbon and oxygen.

$$\overset{\textbf{Required}}{C_{(s)}} + \overset{\textbf{Given}}{O_{2(g)}} \rightarrow CO_{2(g)} + \text{energy}$$

**Step 5.** Find the number of moles of $C_{(s)}$, using the mole ratio in the balanced equation.

$$\frac{\text{Given}}{\text{Required}}\ \frac{1}{1} = \frac{5.00\ \text{mol}\ O_{2(g)}}{X\ \text{mol}\ C_{(s)}} = X = \mathbf{5.00\ mol\ C_{(s)}}$$

**Step 6.** Find the mass of moles of $C_{(s)}$ that can be burnt.

Using the formula, $\mathbf{m = n \times M}$, we get $5.00$ mol $\times 12.01$ g/mol C $= \mathbf{60.10\ g\ C_{(s)}}$

## *Problem:*

7. Sodium azide decomposes extremely quickly to inflate air bags according to the following equation

$$\overset{\textbf{Required}}{2\ NaN_{3(s)}} \rightarrow \overset{\textbf{Given}}{2\ Na_{(s)}} + 3\ N_{2(g)}$$

What mass of sodium azide must decompose to inflate a 70.0 L air bag at SATP?

*Given:*

Volume of $N_{2(g)}$ at SATP: 70.0 L

*Required:*

The mass of sodium azide that must be decomposed to produce 70.0 L of $N_{2(g)}$ at SATP.

*Strategy:*

To solve this problem, convert the volume of $N_{2(g)}$ to amount in moles, then use the mole ratio in the balanced equation to find the number of moles of $NaN_{3(s)}$ that decompose. Knowing the number of moles of $NaN_3$, its mass can be calculated.

**Step 1.** Find the number of moles of nitrogen produced at SATP.

$$n = \frac{70.0\ \text{L}}{24.8\ \text{L/mol}} = \mathbf{2.82\ mol\ N_{2(g)}}\ \text{(1 mole of any gas at SATP is occupies 24.8 L)}$$

**Step 2.** Find the number of moles of $NaN_{3(s)}$ using the mole ratio in the balanced equation.

$$\frac{\text{Given}}{\text{Required}}\ \frac{3}{2} = \frac{2.82\ \text{mol}\ N_{2(g)}}{X\ \text{mol}\ NaN_{3(s)}} = 3X = \mathbf{5.64\ mol\ NaN_{3(s)}}$$

$$X = \mathbf{1.88\ mol\ NaN_{3(s)}}$$

**Step 3.** Find the mass of 1.88 moles of $NaN_{3(s)}$. First find its molar mass.

$M = 1 \times 22.99$ g/mol Na $+ 3 \times 14.01$ g/mol N $= \textit{65.02 g/mol}$

Using the formula, $\mathbf{m = n \times M}$, we get

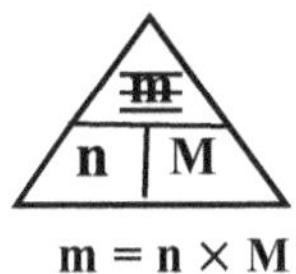


Mass $= 5.64$ mol $\times 65.01$ g/mol $= \mathbf{367\ g\ NaN_{3(s)}}$

## The ideal Gas Equation problems

### *Problem:*

1. Sodium reacts with water to produce hydrogen gas according to the following equation:

**Given**         **Required**

$$2\,Na_{(s)} \;+\; 2\,H_2O_{(l)} \;\rightarrow\; 2\,NaOH_{(aq)} \;+\; H_{2(g)}$$

What volume of $H_{2(g)}$ is produced at 30 °C and 100 kPa if 1.15 g of $Na_{(s)}$ reacts with excess water?

### *Provided:*

Mass of $Na_{(s)}$ = 1.15 g

Temperature of the gas = 30 °C = 303 K

Pressure of the gas = 100 kPa

### *Required:*

The volume of $H_{2(g)}$ produced at 30 °C and 100 kPa if 1.15 g of $Na_{(s)}$ reacts with excess water.

### *Strategy:*

To solve this problem, first find the number of moles of $Na_{(s)}$ in 1.15 g, then use the mole ratio in the balanced equation to find the number of moles of $H_{2(g)}$ produced. Having found the number of moles of $H_{2(g)}$, use the ideal gas equation to find its volume at 100 kPa and 303 K.

**Step 1.** Find the number of moles of $Na_{(s)}$ in 1.15 g.

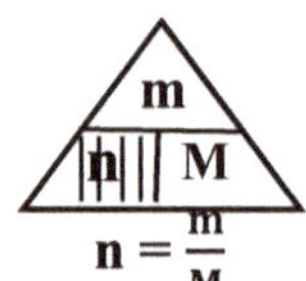

$$\text{Using the equation, } \mathbf{n = \frac{m}{M}}, \text{ we get } \mathbf{n} = \frac{1.15 \text{ g}}{22.99 \text{ g/mol}} = 0.0500 \text{ mol } Na_{(s)}.$$

**Step 2.** Find the number of moles of $H_{2(g)}$ produced

$$\frac{\text{Given}}{\text{Required}}\;\frac{2}{1} = \frac{0.0500 \text{ mol } N_{a(s)}}{X \text{ mol } H_{2(g)}} = 2\,X = 0.0500 \text{ mol } H_{2(g)}$$

$$X = 0.0250 \text{ mol } H_{2(g)}$$

**Step 3.** Find the volume of $H_{2(g)}$ produced by using the ideal gas equation and inserting the values.

Since $PV = nRT$, making V the subject of the equation we get

$$V = \frac{nRT}{P} = \frac{0.0250 \text{ mol} \times \dfrac{8.31 \text{ kPa.L} \times 303 \text{ K}}{\text{mol.K}}}{100 \text{ kPa}} = 0.629 \text{ L} = 629 \text{ mL}$$

### *Problem:*

2. Sodium hydrogen carbonate reacts with hydrochloric acid according to the following equation:

$$NaHCO_{3(s)} \;+\; HCl_{(aq)} \;\rightarrow\; NaCl_{(aq)} \;+\; H_2O_{(l)} \;+\; CO_{2(g)}$$

What volume of $CO_{2(g)}$ is produced at 15 °C and 100 kPa if 3.40 g $NaHCO_{3(s)}$ is allowed to react with 200.0 mL of 0.100 mol/L $HCl_{(aq)}$?

*Provided:*

Mass of $NaHCO_{3(s)}$ = 3.40 g

Volume of $HCl_{(aq)}$ = 200.0 mL = 0.200 L

Concentration of $HCl_{(aq)}$ = 0.100 mol/L

Temperature of $CO_{2(g)}$ = 15 °C = 288 K

Pressure of $CO_{2(g)}$ = 100 kPa

*Strategy:*

This problem entails finding which of the two reagents, $NaHCO_{3(s)}$ or $HCl_{(aq)}$ is the limiting reagent, then using the mole ratio of the balanced equation to find the number of moles of $CO_{2(g)}$. To find the volume of $CO_{2(g)}$, use the ideal gas equation and insert the relevant values.

**Step 1.** Find the number of moles of $HCl_{(aq)}$.

Using the equation, $n = C \times V$, we get

$n$ = 0.100 mol/L $\times$ 0.200 L = **0.0200 mol $HCl_{(aq)}$**

**Step 2.** Find the number of moles of $NaHCO_{3(s)}$. First find its molar mass.

$M$ = 1$\times$ 22.99 g/mol Na + 1$\times$ 1.01 g/mol H+ 1$\times$12.01 g/mol C + 3$\times$ 16.00 g/mol O = **84.01 g/mol**

Using the equation, $n = \dfrac{m}{M}$, we get

$$n = \frac{3.40 \text{ g}}{84.01 \text{ g/mol}} = \textbf{0.0400 mol } NaHCO_{3(s)}\textbf{ (Used)}$$

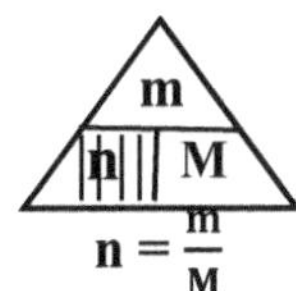


**Step 3.** Find the limiting reagent. Choose 0.0200 mol $HCl_{(aq)}$ to find how much $NaHCO_{3(s)}$ is required with it. The 0.0200 mol $HCl_{(aq)}$ becomes *the given reagent* and the $NaHCO_{3(s}$, *the required one*.

$$\underset{\text{Required}}{NaHCO_{3(s)}} + \underset{\text{Given}}{HCl_{(aq)}} \rightarrow NaCl_{(aq)} + H_2O_{(l)} + CO_{2(g)}$$

$$\frac{\text{Given}}{\text{Required}} \quad \frac{1}{1} = \frac{0.0200 \text{ mol } HCl_{(aq)}}{X \text{ mol } NaHCO_{3(s)}} = X = \textbf{0.0200 mol } NaHCO_{3(s)}\textbf{ (Required)}$$

$$\textbf{0.0400 mol } NaHCO_{3(s)}\textbf{ (Used)}$$

Based on the calculations, it can be concluded that $NaHCO_{3(s)}$ is the excess reagent; **0.0200 mole** was required but only **0.0400 mole** was used. The **0.0200 mole of $HCl_{(aq)}$** is therefore *the limiting reagent* for the next calculation to find how much $CO_{2(g)}$ is produced; it becomes the *given reagent* and $CO_{2(g)}$, *the required* one.

**Step 4.** Use the **0.0200 mole of $HCl_{(aq)}$** to find how much $CO_{2(g)}$ is produced.

$$NaHCO_{3(s)} + \underset{\text{Given}}{HCl_{(aq)}} \rightarrow NaCl_{(aq)} + H_2O_{(l)} + \underset{\text{Required}}{CO_{2(g)}}$$

$$\frac{\text{Given}}{\text{Required}} \quad \frac{1}{1} = \frac{0.0200 \text{ mol } HCl_{(aq)}}{X \text{ mol } CO_{2(g)}} = X = \textbf{0.0200 mol } CO_{2(g)}\textbf{ produced}$$

**Step 5.** Find the volume of $CO_{2(g)}$ produced by using the ideal gas equation and inserting the values.

$$V = \frac{nRT}{P} = \frac{0.0200 \text{ mol} \times \frac{8.31 \text{ kPa.L}}{\text{mol.K}}}{100 \text{ kPa}} \times 288 \text{ } K = \textbf{0.478 L or 478 mL}$$

*Problem:*

**3.** Potassium chlorate decomposes when heated to produce oxygen gas according to the following equation:

$$\overset{\textbf{Required}}{2\ KClO_{3(s)}} \quad \rightarrow \quad \overset{\textbf{Given}}{2\ KCl_{(s)}\ +\ 3O_{2(g)}}$$

What mass of $KClO_{3(s)}$ must be decomposed to produce 50.0 L of $O_{2(g)}$ at 35 °C and 110.0 kPa?

*Provided:*

Volume of $O_{2(g)}$ = 50.0 L

Temperature of $O_{2(g)}$ = 35 °C = 308 K

Pressure of $O_{2(g)}$ = 110.0 kPa

*Strategy:*

To this problem, use the ideal gas equation to find the number of moles of $O_{2(g)}$ and then use the mole ratio in the balanced equation to find the number of moles $KClO_{3(s)}$ decomposed. Knowing this, its mass can be calculated.

**Step 1.** Make the number of moles the subject in the ideal gas equation

Since PV = nRT, $\mathbf{n} = \dfrac{PV}{RT}$. Substituting the relevant values in this equation, we get

$$\mathbf{n} = \frac{110.0\ \text{kPa} \times 50.0\ \text{L}}{\frac{8.31\ \text{kPa.L}}{\text{mol.K}} \times 308\text{K}} = \textbf{2.14 mol } O_{2(g)}$$

**Step 2.** Find the number of moles of $KClO_{3(s)}$.

$$\frac{\text{Given}}{\text{Required}}\ \frac{3}{2} = \frac{2.14\ \text{mol}\ O_{2(g)}}{\text{X mol } KClO_{3(s)}} = 3\ X = \textbf{4.28 mol } KClO_{3(s)}$$

$$\mathbf{X = 1.42\ mol\ } KClO_{3(s)}$$

**Step 3.** Find the mass of $KClO_{3(s)}$. First find its molar mass.

M = 1 × 39.10 g/mol K + 1 × 35.45 g/mol Cl + 3 × 16.00 g/mol O = *122.6 g/mol*

Using the equation, $\mathbf{m = n \times M}$, we get

$$\mathbf{n} = 1.42\ \text{mol} \times 122.6\ \text{g/mol} = \textbf{174 g } KClO_{3(s)}.$$

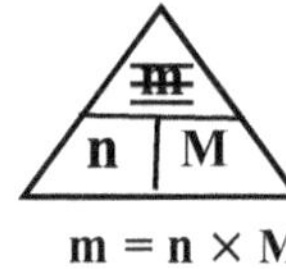


$$m = n \times M$$

*Problem:*

**4.** Magnesium reacts with hydrochloric acid according to the following balanced equation:

$$\overset{\textbf{Given}}{Mg_{(s)}} \quad + \quad 2\ HCl_{(aq)} \quad \rightarrow \quad MgCl_{2(aq)}\ +\ H_{2(g)}$$

What volume of $H_{2(g)}$ is produced at 40 °C and 105.0 kPa if 1.20 g of $Mg_{(s)}$ reacts with 400.0 mL of 1.00 mol/L $HCl_{(aq)}$?

*Provided:*

Mass of Mg = 1.20 g

Volume of $HCl_{(aq)}$ 400.0 mL = 0.400 L

Concentration of $HCl_{(aq)}$ = 1.00 mol/L

Temperature of $H_{2(g)}$ = 40 °C = 313 K

Pressure of $H_{2(g)}$ = 105.0 kPa

*Strategy:*

This problem entails finding which of the two reagents, $Mg_{(s)}$ or $HCl_{(aq)}$ is the limiting reagent, then using the mole ratio of the balanced equation to find the number of moles of $H_{2(g)}$ produced. To find the volume of $H_{2(g)}$ produced, use the ideal gas equation and insert the relevant values.

**Step 1.** Find the number of moles of $HCl_{(aq)}$.

Using the equation, $\mathbf{n} = \mathbf{C} \times \mathbf{V}$, we get

$$\mathbf{n} = 0.100 \text{ mol/L} \times 0.400 \text{ L} = \mathbf{0.0400 \text{ mol } HCl_{(aq)} \text{ (Used)}}$$

**Step 2.** Find the number of moles of $Mg_{(s)}$.

Using the equation, $\mathbf{n} = \dfrac{\mathbf{m}}{\mathbf{M}}$, we get $\mathbf{n} = \dfrac{1.20 \text{ g}}{24.31 \text{ g/mol}}$

$$= \mathbf{0.050 \text{ mol } Mg_{(s)}}$$

**Step 3.** Find the limiting reagent. Choose the **0.500 mol $Mg_{(s)}$** to find how much $HCl_{(aq)}$ is required to react with it. $Mg_{(s)}$ become the *given reagent* and $HCl_{(aq)}$, *the required one*.

**Given**     **Required**

$Mg_{(s)}$ +   2 $HCl_{(aq)}$   $\rightarrow$   $MgCl_{2(aq)}$ + $H_{2(g)}$

$$\frac{\text{Given}}{\text{Required}} \quad \frac{1}{2} = \frac{0.050 \text{ mol } Mg_{(s)}}{\text{X mol } HCl_{(aq)}} = X = \mathbf{0.100 \text{ mol } HCl_{(aq)} \text{ (Required)}}$$

$$\mathbf{0.040 \text{ mol } HCl_{(aq)} \text{ (Used)}}$$

Comparing these two values, it can be seen that less $HCl_{(aq)}$ was used (**0.040 mol**) than was required (**0.100 mol**) to react with the $Mg_{(s)}$; it therefore is *the limiting reagent*. For the next calculation, it becomes the *given reagent* and the $H_{2(g)}$ becomes *the required one*.

**Step 4.** Find the number of moles of $H_{2(g)}$ produced.

**Given**            **Required**

$Mg_{(s)}$ +   2 $HCl_{(aq)}$   $\rightarrow$   $MgCl_{2(aq)}$ + $H_{2(g)}$

$$\frac{\text{Given}}{\text{Required}} \quad \frac{2}{1} = \frac{0.040 \text{ mol } HCl_{(aq)}}{\text{X mol } H_{2(g)}} = 2 \, X = 0.040 \text{ mol } H_{2(g)}$$

$$\mathbf{X = 0.020 \text{ mol } H_{2(g)}}$$

**Step 5.** Find the volume of 0.020 mol $H_{2(g)}$ at 40 °C and 105.0 kPa, using the ideal gas equation.

$$V = \frac{nRT}{P} = \frac{0.020 \text{ mol} \times \dfrac{8.31 \text{ kPa.L} \times 313 \text{ K}}{\text{mol.K}}}{105.0 \text{ kPa}} = \mathbf{0.495 \text{ L } H_{2(g)}}$$

## *Problem:*

**5.** What volume of $H_{2(g)}$ is produced at 30 °C and 100.0 kPa between the reaction of 300.0 mL of 1.0 mol/L $H_2SO_{4(aq)}$ and 2.4 g of $Mg_{(s)}$.

*Provided:*

Mass of Mg = 2.4 g

Volume of $H_2SO_{4(aq)}$ = 300.0 mL = 0.300 L

Concentration of $H_2SO_{4(aq)}$ = 1.00 mol/L

Temperature of $H_{2(g)}$ = 30 °C = 303 K

Pressure of $H_{2(g)}$ = 100.0 kPa

This problem entails finding which of the two reagents, $Mg_{(s)}$ or $H_2SO_{4(aq)}$ is the limiting reagent, then using the mole ratio of the balanced equation to find the number of moles of $H_{2(g)}$ produced. To find the volume of $H_{2(g)}$ produced, use the ideal gas equation and insert the relevant values.

**Step 1.** Find the number of moles of $H_2SO_{4(aq)}$

Using the equation, $\mathbf{n = C \times V}$, we get

$\mathbf{n}$ = 0.100 mol/L $\times$ 0.300 L = **0.0300 mol $H_2SO_{4(aq)}$ (Used)**

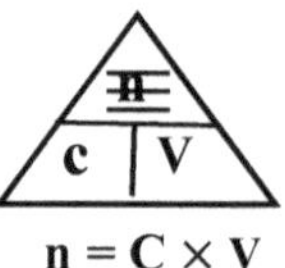


$\mathbf{n = C \times V}$

**Step 2.** Find the number of moles of $Mg_{(s)}$.

Using the equation, $\mathbf{n = \dfrac{m}{M}}$, we get $n = \dfrac{2.40\ g}{24.31\ g/mol}$



$n = \dfrac{m}{M}$

= **0.100 mol $Mg_{(s)}$**

**Step 3.** Find the limiting reagent. Choose the **0.100 mol $Mg_{(s)}$** to find how much $H_2SO_{4(aq)}$ is required to react with it. $Mg_{(s)}$ becomes the *given reagent* and $H_2SO_{4(aq)}$, the *required one.*

$$\begin{array}{ccccccc} \textbf{Given} & & \textbf{Required} & & & & \\ Mg_{(s)} & + & H_2SO_{4(aq)} & \rightarrow & MgSO_{4(aq)} & + & H_{2(g)} \end{array}$$

$$\frac{\text{Given}}{\text{Required}} \ \frac{1}{1} = \frac{0.100\ mol\ Mg_{(s)}}{X\ mol\ H_2SO_{4(aq)}} = X = \textbf{0.100 mol } H_2SO_{4(aq)} \textbf{ (Required)}$$

**0.0300 mol $H_2SO_{4(aq)}$ (Used)**

Comparing these two values, it can be seen that less $H_2SO_{4(aq)}$ was used (**0.0300 mol**) than was required (**0.100 mol**) to react with the $Mg_{(s)}$; it therefore is *the limiting reagent*. For the next calculation, it becomes the *given reagent* and the $H_{2(g)}$ becomes *the required one*.

**Step 4.** Find the number of moles of $H_{2(g)}$ produced.

$$\begin{array}{ccccccc} & & \textbf{Given} & & & & \textbf{Required} \\ Mg_{(s)} & + & H_2SO_{4(aq)} & \rightarrow & MgSO_{4(aq)} & + & H_{2(g)} \end{array}$$

$$\frac{\text{Given}}{\text{Required}} \ \frac{1}{1} = \frac{0.0300\ mol\ H_2SO_{4(aq)}}{X\ mol\ H_{2(g)}} = X = \textbf{0.0300 mol } H_{2(g)}$$

**Step 5.** Find the volume of 0.0300 mol $H_{2(g)}$ at 30 °C and 100.0 kPa, using the ideal gas equation.

$$V = \frac{nRT}{P} = \frac{0.0300\ mol \times \dfrac{8.31\ kPa.L \times 303\ K}{mol.K}}{100.0\ kPa} = \textbf{0.755 L } H_{2(g)}$$

## Problem:

**6.** What mass of of $Zn_{(s)}$ must have been added to 200.0 mL of 0.10 mol/L $H_2SO_{4(aq)}$ to produce 248.0 mL of $H_{2(g)}$ at 25 °C and 100.0 kPa?

*Provided:*

Volume of $H_2SO_{4(aq)}$ = 200.0 mL = 0.200 L

Concentration of $H_2SO_{4(aq)}$ = 0.10 mol/L

Volume of $H_{2(g)}$ = 248.0 mL = 0.2480 L

Temperature of $H_{2(g)}$ = 25 °C = 298 K

Pressure of $H_{2(g)}$ = 100.0 kPa

*Strategy:*

This problem entails first, finding the number of moles of $H_{2(g)}$ produced, using the ideal gas equation, and then using that to find the number of moles of $Zn_{(s)}$ required to produce it when reacting with $H_2SO_{4(aq)}$. In this case, we shall assume that the amount of $H_2SO_{4(aq)}$ used is in excess to allow the required amount $Zn_{(s)}$ to be react.

*Steps to be followed:*

**Step 1.** Use the ideal gas equation and make number of moles, **n** the subject of the equation

$$n = \frac{P \times V}{RT} = n = \frac{100.0 \text{ kPa} \times 0.248 \text{ L}}{\frac{8.31 \text{ kPa.L}}{\text{mol.K}} \times 2\,98 \text{ K}} = \textbf{0.0100 mol } H_{2(g)}$$

**Step 2.** Use the **0.0100 mol $H_{2(g)}$** produced, to find the number of moles of $Zn_{(s)}$ used.

| Required | | | | Given |
|---|---|---|---|---|
| Zn | + | $H_2SO_{4(aq)}$ | $\rightarrow$ ZnSO$_{4(aq)}$ + | $H_{2(g)}$ |

$$\frac{\text{Given}}{\text{Required}} \quad \frac{1}{1} = \frac{0.0100 \text{ mol } H_{2(g)}}{X \text{ mol } Zn_{(s)}} = X = \textbf{0.0100 mol } Zn_{(s)}$$

**Step 3.** Find the mass of 0.0100 mol $Zn_{(s)}$.

Using the equation, $\mathbf{m = n \times M}$ we get

Mass = 0.0100 mol $\times$ 65.39 g/mol = **0.654 g $Zn_{(s)}$**

*Confirm that the $H_2SO_{4(aq)}$ used is in excess.*

$$m = n \times M$$

**Step 4.** Find the number of moles of $H_2SO_{4(aq)}$ used.

Using the equation, $\mathbf{n = C \times V}$, we get

$\mathbf{n = 0.100 \text{ mol/L} \times 0.200 \text{ L} = 0.0200 \text{ mol } H_2SO_{4(aq)} \text{ (Used)}}$

**Step 5.** Find the number of moles of $H_2SO_{4(aq)}$ required to react with 0.0100 mole of $Zn_{(s)}$.

In this case, the $Zn_{(s)}$ becomes the *given reagent* and the $H_2SO_{4(aq)}$, the *required one*.

$$\frac{\text{Given}}{\text{Required}} \quad \frac{1}{1} = \frac{0.0100 \text{ mol } Zn_{(s)}}{X \text{ mol } H_2SO_{4(aq)}} = X = \textbf{0.0100 mol } H_2SO_{4(aq)} \textbf{ (Required)}$$

$$\textbf{0.0200 mol } H_2SO_{4(aq)} \textbf{ (Used)}$$

It can be confirmed that the amount of $H_2SO_{4(aq)}$ was indeed in excess; more was used (**0.0200 mol**) than was required)

## *Problem:*

7. 50.0 mL of 0.40 mol/L of a certain acid was reacted with excess magnesium and 258.4 mL of hydrogen was produced at 38 °C and 100 kPa. Was the acid used $HCl_{(aq)}$ or $H_2SO_{4(aq)}$?

*Strategy:*

This problem entails first, finding the number of moles of $H_{2(g)}$ produced using the ideal gas equation. Then, use the number of moles of $H_{2(g)}$ produced to find the number of moles of each of the acids required to produce it, using balanced equations between magnesium and both two acids. Finally, find the number of moles for each of the two acids, using their volumes and concentrations provided, **and see which one corresponds to the correct one required.**

*Steps to be carried out:*

**Step 1.** Find the number of moles of moles of $H_{2(g)}$ produced. Making **n** the subject of the ideal gas equation, we get $n = \dfrac{PV}{RT}$. Substituting the relevant values in this equation, we get

$$n = \frac{100.0 \text{ kPa} \times 0.2584 \text{ L}}{\frac{8.31 \text{ kPa.L}}{\text{mol.K}} \times 311 \text{K}} = \textbf{0.0100 mol } H_{2(g)}$$

**Step 2.** Write a balanced equation between Mg and $H_2SO_{4(aq)}$ to find how many moles of the $H_2SO_{4(aq)}$ that are required to produce 0.0100 mol $H_{2(g)}$.

        **Required**                 **Given**

$Mg_{(s)} \quad + \quad H_2SO_{4(aq)} \quad \rightarrow \quad MgSO_{4(aq)} \quad + \quad H_{2(g)}$

**Step 3.** Find the number of moles of $H_2SO_{4(aq)}$ that can produce 0.0100 mol $H_{2(g)}$.

$$\frac{\text{Given}}{\text{Required}} \quad \frac{1}{1} = \frac{0.0100 \text{ mol } H_{2(g)}}{X \text{ mol } H_2SO_{4(aq)}} = \boxed{X = 0.0100 \text{ mol } H_2SO_{4(aq)}}$$

**Step 4.** Write a balanced equation between Mg and $HCl_{(aq)}$ to find how many moles of the $HCl_{(aq)}$ are required to produce 0.0100 mol $H_{2(g)}$.

        **Required**              **Given**

$Mg_{(s)} \quad + \quad 2 \, HCl_{(aq)} \quad \rightarrow \quad MgCl_{2(aq)} \quad + \quad H_{2(g)}$

**Step 5.** Find the number of moles of $HCl_{(aq)}$ that can produce 0.0100 mol $H_{2(g)}$.

$$\frac{\text{Given}}{\text{Required}} \quad \frac{1}{2} = \frac{0.0100 \text{ mol } H_{2(g)}}{X \text{ mol } HCl_{(aq)}} = \boxed{X = 0.0200 \text{ mol } HCl_{(aq)}}$$

**Step 6.** Find which of the two acids was used based on the data provided.

- If it were **$H_2SO_{4(aq)}$** the number of moles it will provide is found by

  Using the equation, **n = C × V**

  $$n = 0.400 \text{ mol/L} \times 0.0500 \text{ L} = \textbf{0.0200 mol}$$

- If it were $HCl_{(aq)}$ the number of moles it will provide is found by

  Using the equation, **n = C × V**

  $$n = 0.400 \text{ mol/L} \times 0.0500 \text{ L} = \boxed{\textbf{0.0200 mol}}$$

**n = C × V**

Based on the calculations, it can be concluded that **the 50.0 mL** of **0.400 mol/L $HCl_{(aq)}$** *provides the correct amount* to react with excess $Mg_{(s)}$ to produce 0.0100 mol $H_{2(g)}$. Using the $H_2SO_{4(aq)}$ would provide an excess of what was required **(0.0200 mol),** whereas only **0.0100 mol** was required.

**Exercise 20.11 (p.356)**

## The Nitrogen Cycle

**1.** Two nitrogen atoms are triply bonded to form a molecule of nitrogen. How might the percentage composition of nitrogen in the atmosphere change if the atoms were doubly bonded?

The percentage of composition of nitrogen in the atmosphere would decrease. The strength of a nitrogen double bond is much weaker than its current triple bond which makes the molecule very stable. Double-bonded nitrogen molecules would be much easier to break and that would allow nitrogen to react more readily with other elements, thereby decreasing its percentage composition in the atmosphere.

186

**2.** At the end of a growing season, lots of nutrients are still locked up in the leaves, roots and stems of plants. Instead of burning or discarding these as wastes, why would it be more beneficial to the soil if these are allowed to decompose in it?

Allowing the dead plants to stay and decompose on the soil allows for the locked-up minerals to be released from the decomposed tissues into the soil. This would ensure that most of the minerals taken up by the plants from the soil are returned to it. Discarding these away from the sites, is counter productive as it would not allow for the nutrients to replenish the depleted soil.

**3.** During the denitrification process, **anaerobic** bacteria utilize oxygen bonded to nitrate ions for their respiration. In so doing, the bacteria decompose these valuable ions present in the soil, producing nitrogen gas. What must farmers do to ensure that very little amount of nitrate is lost from the soil?

Farmers must ensure that the soil always remains well aerated to prevent anaerobic conditions. For farming done in temperate countries, this should be done especially at the beginning of Spring, when the degree of aeration is not good due to waterlogging from Winter.

**4.** Fish farmers use nitrogenous fertilizers in water to promote the growth of algae that the fish feed on. Why must care be taken not to over fertilize the water?

If they over fertilize, it will cause algal blooms with their subsequent death. The dead algae increase the biological demand of the water due to the proliferation and activities of decomposing bacteria. The increased bacterial activity reduces dissolved oxygen that is essential for the fish's survival. Fish death and subsequent anaerobic conditions may result.

# Chapter Review:  Gas Mixture (18-20)

## Matching

*Match each term in the table below with the correct statements that follow*

| | | | |
|---|---|---|---|
| A | Dalton | G | Gay-Lussac |
| B | Charles | H | Kelvin |
| C | $P_1V_1 = P_2V_2$ | I | 100 kPa and 25°C |
| D | Boyle | J | $V = kT$ |
| E | $P = kT$ | K | $PV = nRT$ |
| F | Avogadro | L | 24.8 L |

| | |
|---|---|
| **1.** | He formulated the law of partial pressures for gases. |
| **2.** | He extrapolated graphs of volume versus temperature for real gases to find absolute zero. |
| **3.** | He discovered that volume of a fixed mass of gas varies directly with temperature changes on the gas. |
| **4.** | He hypothesized that equal volumes of gases at the same pressure and temperature contain the same number of particles. |
| **5.** | He formulated the law of combining volumes for gaseous reactants and products during reactions of gases. |
| **6 .** | He formulated law that the volume of a gas is inversely proportional to the pressure acting upon it. |
| **7.** | Boyle's law equation. |
| **8.** | Standard ambient pressure and temperature. |
| **9.** | Molar volume at SATP. |
| **10.** | Charles' law equation. |
| **11.** | The pressure law equation. |
| **12.** | The ideal gas equation. |

## True or False

**Read each of the following statements and then decide if it is *True* or *False***

| | |
|---|---|
| 1. | The pressure of a gas in a container is due to the bombardments of the gas particles against its walls. |
| 2. | The particles in an ideal gas attract each other and the walls of the container they bombard. |
| 3. | As the temperature of a gas is raised so does the average kinetic energy of its particles. |
| 4. | The higher we ascend into the atmosphere, the greater the atmospheric pressure becomes. |

| 5. | For a fixed mass of gas at constant volume, its pressure decreases as its temperature is raised. |
|---|---|
| 6. | Increasing the number of gas particles in a container at constant temperature, raises its pressure. |
| 7. | The volume of a fixed mass of gas at constant pressure will increase if its temperature is raised. |
| 8. | On top of a high mountain, water may boil as low as 70 $^{\circ}$C |
| 9. | The pressure of a fixed mass of gas of fixed volume will increase if its temperature is lowered. |
| 10. | The total pressure of a mixture of non-reacting gases is due to the sum of the partial pressures of each gas making up the mixture. |
| 11. | One mole of $H_{2(g)}$ occupies the same volume as one mole $CO_{2(g)}$ at SATP. |
| 12. | 16.0 g of $O_{2(g)}$ will have the same volume as 1.0 g $H_{2(g)}$ at SATP. |

## Multiple Choice

**Choose the letter that best answers questions.**

1.

| Which of the following are **true** statements about real gases? | |
|---|---|
| a | They are highly compressible |
| b | Their particles attract each other upon collision |
| c | They condense when they are cooled to low temperatures |
| d | The force of attraction between their particles is very weak |
| e | All of the above |

2.

| Which of the following statements is **false** about ideal gases? | |
|---|---|
| a | Their particles have negligible volumes |
| b | Their particles are not attracted to the walls of their container |
| c | They do not condense when they are cooled to low temperatures |
| d | There is no force of attraction between their particles. |
| e | none of the above |

3.

| The pressure of the gas in an open container at sea level is | |
|---|---|
| a | the same as that of the atmosphere |
| b | different from that of the atmosphere |
| c | approximately 101 kPa |
| d | a and c only |
| e | none of the above |

4.

|   | Atmospheric pressure is caused by |
|---|---|
| a | the planets |
| b | the force of gravity on the air particles above the Earth's surface |
| c | the various things on the Earth's surface |
| d | the wind blowing over the Earth's surface |
| e | none of the above |

5.

|   | If a mixture of hydrogen gas and chlorine gas react in a sealed container, there will be |
|---|---|
| a | a decrease of pressure |
| b | no difference in pressure when the products cool down |
| c | an initial increase in pressure |
| d | all of the above |
| e | b and c |

6.

| The pressure of a gas in a container at a fixed temperature can be increased by | | | |
|---|---|---|---|
| a | reducing it volume | d | reducing the number of particles |
| b | increasing its volume | e | a and c only |
| c | increasing the number of particles | | |

7.

| A fixed mass of gas has a volume of 2.0 L at 100 kPa. Its volume at 200.0 kPa will be | | | |
|---|---|---|---|
| a | 2.0 L | d | 4.0 L |
| b | 1.0 L | e | none of the above |
| c | 20.0 L | | |

8.

| A fixed mass of gas has a volume of 2.0 L at 100 kPa. At what pressure will its volume be 0.5 L? | | | |
|---|---|---|---|
| a | 200 kPa | d | 400 kPa |
| b | 300 kPa | e | none of the above |
| c | 50 kPa | | |

9.

| The pressure of a fixed mass of gas at constant volume increases with temperature because its | | | |
|---|---|---|---|
| a | molecules break up | d | molecules collide more frequently |
| b | molecules gain more energy | e | all except a |
| c | molecules move more forcefully | | |

10.

A rigid container has a mixture of $N_{2(g)}$, $CO_{2(g)}$ and $O_{2(g)}$ in a 12: 5: 3 mole ratios respectively.

If the total pressure in the container is 100.0 kPa, the partial pressure of $N_{2(g)}$ must be

| | | | |
|---|---|---|---|
| a | 60.0 kPa | d | 12.0 kPa |
| b | 5.0 kPa | e | 25.0 kPa |
| c | 15.0 kPa | | |

11.

A rigid container has 10.0 L of $NH_{3(g)}$ at SATP. What volume of $H_{2(g)}$ can be collected at SATP if all the $NH_{3(g)}$ is decomposed?

| | | | |
|---|---|---|---|
| a | 10.0 L | d | 15.0 L |
| b | 20.0 L | e | none of the above |
| c | 30.0 L | | |

12.

The pressure of oxygen gas collected above water at a certain temperature is 101.0 kPa. If the vapor pressure of water at that temperature is 4.0 kPa, the pressure of dry oxygen gas will be?

| | | | |
|---|---|---|---|
| a | 101.0 kPa | d | 115.0 kPa |
| b | 97.0 kPa | e | none of the above |
| c | 4.0 kPa | | |

13.

Sulphur dioxide and oxygen react according to the following equation:

$$2\ SO_{2(g)} + O_{2(g)} \rightarrow 2\ SO_{3(g)}$$

Which of the following ratios represents the correct volumes of reactants and products?

| | | | |
|---|---|---|---|
| a | 1: 2: 1 | d | 2: 1: 1 |
| b | 1: 1: 1 | e | none of the above |
| c | 2: 1: 2 | | |

14.

If one mole of any gas at STP occupies 22.4 L, the volume occupied by 1.01 g $H_{2(g)}$ will be

| | | | |
|---|---|---|---|
| a | 1.01 L | d | 2.02 L |
| b | 11.2 L | e | none of the above |
| c | 22.8 L | | |

15.

At STP 8.0 g of $O_{2(g)}$ and 7.0 g of $N_{2(g)}$ will occupy

| | | | |
|---|---|---|---|
| a | the same volume | d | 11.2 L each |
| b | different volumes | e | a and c only |
| c | 5.6 L each | | |

16.

The number of molecules in 6.2 L of $CO_{2(g)}$ at SATP is

| | | | | |
|---|---|---|---|---|
| *a* | $1.50 \times 10^{23}$ | *d* | $6.02 \times 10^{23}$ |
| *b* | $3.01 \times 10^{23}$ | *e* | none of the above |
| *c* | $1.50 \times 10^{24}$ | | |

17.

The following equation represents the reaction between magnesium and sulphuric acid:

$Mg_{(s)} + H_2SO_{4(aq)} \rightarrow MgSO_{4(aq)} + H_{2(g)}$. If 2.0 g of $Mg_{(s)}$ reacts with excess $H_2SO_{4(aq)}$, what volume of $H_{2(g)}$ at STP will be produced?

| | | | | |
|---|---|---|---|---|
| *a* | 0.92 L | *d* | 3.68 L |
| *b* | 1.84 L | *e* | none of the above |
| *c* | 44.8 L | | |

18.

The following equation represents the reaction between hydrochloric acid and sodium carbonate: $2\ HCl_{(aq)} + Na_2CO_{3(s)} \rightarrow 2\ NaCl_{(aq)} + CO_{2(g)} + H_2O_{(l)}$. What mass of $Na_2CO_{3(s)}$ is required to react with excess $HCl_{(aq)}$ to produce 12.4 L of $CO_{2(g)}$ at SATP?

| | | | | |
|---|---|---|---|---|
| *a* | 105.99 g | *d* | 211.98 g |
| *b* | 52.99 g | *e* | none of the above |
| *c* | 26.50 g | | |

19.

If a rigid 20.0 L container has 64.2 g $CH_{4(g)}$, what is the pressure of the gas if the cylinder is kept at a temperature of 30 °C?

| | | | | |
|---|---|---|---|---|
| *a* | 101 kPa | *d* | 504 kPa |
| *b* | 100 kPa | *e* | none of the above |
| *c* | 250 kPa | | |

20.

Which of the following gases cause destruction of the ozone layer?

| | | | | |
|---|---|---|---|---|
| *a* | $CO_2$ | *d* | CO |
| *b* | CFC's | *e* | $CH_4$ |
| *c* | $SO_2$ | | |

21.

Decomposition of potassium chlorate produces oxygen as follows:

$$2\ KClO_{3(s)} \rightarrow 2\ KCl_{(s)} + 3\ O_{2(g)}$$

What mass of $KClO_{3(s)}$ must be decomposed to produce 50 L of $O_{2(g)}$ at 250 °C and 120 kP?

| | | | | |
|---|---|---|---|---|
| *a* | 112.7 g | *d* | 245.1 g |
| *b* | 122.55 g | *e* | none of the above |
| *c* | 81.7 g | | |

**22.**

Which of the following gases **does not** cause global warming?

| | | | |
|---|---|---|---|
| a | $CO_2$ | d | CO |
| b | NO | e | $CH_4$ |
| c | $H_2O_{(g)}$ | | |

**23.**

Which of the following gases causes blood poisoning?

| | | | |
|---|---|---|---|
| a | $CO_2$ | d | $SO_2$ |
| b | CFC's | e | $CH_4$ |
| c | CO | | |

**24.**

Which of the following gases **would not cause** acid rain?

| | | | |
|---|---|---|---|
| a | $NO_2$ | d | $SO_2$ |
| b | CO | e | $SO_3$ |
| c | $CO_2$ | | |

**25.**

Which of the following gases is found in rotten eggs?

| | | | |
|---|---|---|---|
| a | $NO_2$ | d | $H_2S$ |
| b | $CH_4$ | e | $SO_2$ |
| c | $CO_2$ | | |

**26.**

This greenhouse gas is found in cow's flatulence?

| | | | |
|---|---|---|---|
| a | $NO_2$ | d | $CO_2$ |
| b | $CH_4$ | e | $H_2O$ |
| c | $SO_2$ | | |

# Chapter Review: Gases (18-20) Solutions

**Matching :**

| 1.) A | 2.) H | 3.) B | 4.) F | 5.) G | 6.) D | 7.) C | 8.) I | 9.) L |
|---|---|---|---|---|---|---|---|---|
| 10.) J | 11.) E | 12.) K | | | | | | |

**True/False**

| 1.) T | 2.) F | 3.) T | 4.) F | 5.) F | 6.) T | 7.) T | 8.) T | 9.) F |
|---|---|---|---|---|---|---|---|---|
| 10.) T | 11.) T | 12.) T | | | | | | |

**Multiple Choice:**

| 1.) E | 2.) E | 3.) D | 4.) B | 5.) E | 6.) E | 7.) B | 8.) D | 9.) E |
|---|---|---|---|---|---|---|---|---|
| 10.) A | 11.) D | 12.) B | 13.) C | 14.) B | 15.) E | 16.) A | 17.) B | 18.) B |
| 19.) D | 20.) E | 21.) A | 22.) D | 23.) C | 24.) B | 25.) D | 26.) B | |

# The Periodic Table of the Elements

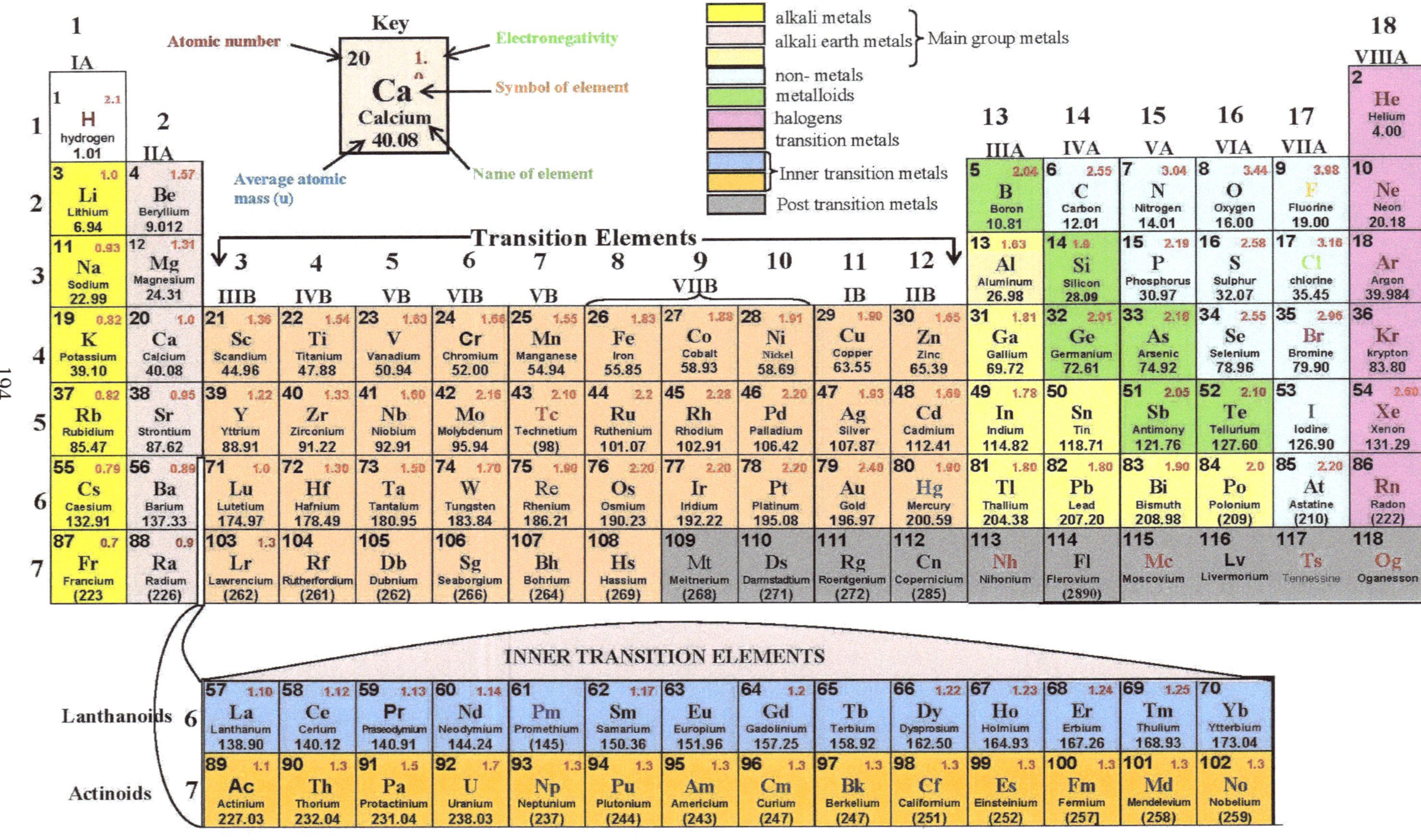